W9-BGW-485

FLORIDA STATE
UNIVERSITY LIBRARIES

OCT 15 1999

TALLAHASSEE, FLORIDA

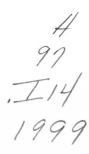

Published 1999 by Humanity Books, an imprint of Prometheus Books

Philosophical Ecologies: Essays in Philosophy, Ecology, and Human Life. Copyright © 1999 A. Pablo Iannone. All rights reserved. No part of this publication may be reproduced, stored in a retrieval system, or transmitted in any form or by any means, electronic, mechanical, photocopying, recording, or otherwise, without prior written permission of the publisher, except in the case of brief quotations embodied in critical articles and reviews. Inquiries should be addressed to Humanity Books, 59 John Glenn Drive, Amherst, New York 14228–2197, 716–691–0133, ext. 207. FAX: 716–564–2711.

03 02 01 00 99   5 4 3 2 1

Library of Congress Cataloging-in-Publication Data

Iannone, A. Pablo.
    Philosophical ecologies : essays in philosophy, ecology, and human life / A. Pablo Iannone.
            p.   cm.
    Includes bibliographical references and index.
    ISBN 1–57392–660–4 (cloth : alk. paper)
    1. Policy sciences—Philosophy. 2. Social ecology—Philosophy. 3. Social problems. I. Title.
H97.I14   1999
300—dc21   99–24179
                                                                             CIP

Printed in the United States of America on acid-free paper

To my wife, Mary Kay Garrow, whose wisdom and experience helped give concrete meaning to this book's ideas, and whose contributions in writing essay five helped make this book much better than it would have been otherwise. and to our daughters, Alejandra Emilia and Catalina Patricia, whose joy of life, love of learning, and growing search for their place in the universe helped give concrete meaning to this book's aspirations.

Contents

An ethic, ecologically, is a limitation on freedom of action in the struggle for existence. An ethic, philosophically, is a differentiation of social from anti-social conduct. These are two definitions of one thing. The thing has its origin in the tendency of interdependent individuals or groups to evolve modes of cooperation. The ecologist calls these symbioses. Politics and economics are advanced symbioses in which the original free-for-all competition has been replaced, in part, by cooperative mechanisms with an ethical content.

—Aldo Leopold, *A Sand County Almanac and Sketches Here and There*

Preface

Our times, like all times, are not easy times in which to live. They are, however, especially complicated because of the unprecedented fragmentation they involve between business technology and policy-making sectors, policy-making sectors and the general public; one artistic tradition and another; and the myriad of cultures that are coming into increasing—sometimes conflictive—contact during the later part of the twentieth century. This book grew from the realization that such fragmentation constitutes a significant obstacle to a range of human activities that are crucial to any human flourishing. For, first, it is a significant obstacle to politically sound and morally sensitive policy making. Second, it is an equally significant obstacle to any reasonably harmonious interaction between different groups, between groups and their members, and between individuals. Third, this fragmentation also undermines any harmonious interactions between humans and their natural environment. Fourth, especially in cross-cultural interactions, it undermines the flourishing of ethnic groups and their members, when these are—but sometimes also when they are not—disadvantaged.

This book is designed to help correct the situation beyond the scope of policy making, into the areas of social conventions and practices, and personal, moral, intellectual, and aesthetic development. It builds on work carried out in my previous books: *Contemporary Moral Controversies in Technology* (New York and London: Oxford University Press, 1987), *Contemporary Moral Controversies in Business* (New York and London: Oxford University Press, 1989), *Through Time and Culture* (Englewood Cliffs, N.J.: Prentice Hall, 1994) and, most important, *Philosophy as Diplomacy* (Amherst, N.Y.: Humanity Books, 1994). These provided evidence for the said lack of integration and began to address the problems it poses. *Philosophy as Diplomacy* is salient among these works. It thoroughly formulated a theoretical framework that combines rights, consequences, and pragmatic considerations. It also developed a taxonomy of policymaking situations and described various social decision procedures applicable to them. Finally, it used the framework, taxonomy, and procedures thus developed to examine a great variety of policy-making cases.

These results are substantially expanded and extended to other areas in the book. It moves beyond policy making and pays much attention to convention- and practice-settling processes, interactions, and mutual adaptations as they apply to intercultural conflicts, environmental issues, and tensions between personal values and organizational behavior. In focusing on these processes, interactions, and adaptations, the book's approach is significantly analogous to approaches found in ecological studies, especially those in the branch of sociology known as *ecology*, which is "concerned with the spacing of people and institutions and the resulting, interdependency."[1] Hence its title, *Philosophical Ecologies*. The book breaks new ground in applying an ecological model to a wide range of philosophical problems and issues. Some are environmental, others intercultural, still others about matters of aesthetics and the place and role of science, ideology, and philosophy in our fragmented world.

The book proceeds in accordance with two research methods. One is taxonomic and has been used in both philosophy and social process studies. It consists in merging description and argument. That is, it describes a number of cases or observations to exemplify categories, classifications, and theories and then uses these (or similar) cases or observations as evidence for the utility or accuracy of the categories, classifications, and theories. This method is currently being used in social science to study such issues as contravention and conflict in administrative entities.[2]

This taxonomic method is also in current use among philosophers who attempt to connect conceptual analysis or phenomenological studies with the results of scientific research, the experiences of the general public, and the aims of ethics and sociopolitical philosophy. A relatively recent work influenced by the analytic tradition, which exemplifies this approach, is David Braybrooke 's *Meeting Needs*.[3] In addition, I significantly used this method in *Philosophy as Diplomacy*.

The other, ecological method was introduced by Edith Cobb in *The Ecology of Imagination in Childhood*[4] (New York: Columbia University Press, 1977). Cobb used this method, which is based on the ecological model of interchanges and mutual adaptations and interconnections between organisms and their environment, to analyze children's development of cosmic sense. As Margaret Mead commented, however, the method has a wide scope of applicability and "makes it possible to compare the most diverse cultural developments."[5]

This ecological method, for example, may help explain not only cross-cultural interactions such as the mutual adaptations the expatriate experience involves between individuals and cultures other than their own, but also the adaptations developed between groups. *Philosophical Ecologies* will apply concepts of ecology to the study of a wide variety of activities that pose philosophical problems and social issues.

This book is of interest to the initiated and accessible to the uninitiated.

Philosophical language and categories are introduced by reference to problems raised, and language and categories are used in discussions of cases by policymakers and ordinary people. This helps meet the needs of those more familiar with the world of political, economic, and other everyday decisions than with that of philosophy. It also helps clarify the extent to which particular positions in moral philosophy have practical applications in the actual policy making and the more general social world.

The book is unique or nearly unique. For its treatment of theoretical and practical problems is guided by two concerns crucial to the integrated approach it develops. The first is a concern with traditional and contemporary theories in ethics, social and political philosophy, and other branches of philosophy. And the second is a concern with the applicability of all these theories to actual policy-making practice and, in general, practical situations faced by humans as individuals and as groups.

I know of no other book that does this with the degree of detail and reliance on empirical information one finds in this book. For, first, to the best of my knowledge, the books being published in the area of philosophy and policy making, and philosophy and social problems/issues, presuppose somewhat simplistic conceptions of policy making and social problems/issues. Second, they do not rely on empirical social studies to the extent this book does. Third, though there are many philosophical discussions of environmental issues that use the scientific results of ecological studies to support philosophical positions about the issues, they do not go much—if any—further than that. By contrast, this book applies ecological concepts to the study of the issues themselves, including the interactions between humans, as well as between humans and nonhumans, involved in the issues. In this regard, the book moves into largely uncharted—but quite promising—territory. Fourth, in contrast to this book, those being published in the areas of philosophy and policy making and philosophy and social problems/issues primarily, if not exclusively, use philosophical language and categories. Hence, as indicated, they make it difficult, if not impossible, to establish the extent to which the particular positions they take have practical applications in the actual world.

Philosophical Ecologies includes twelve essays interconnected by dialogues and divided into four parts, a selected bibliography, and an index. The essays are self-contained. Yet, taken together, they carry out the book's two primary tasks. The first is largely taxonomic, involving the previously mentioned characterization and classification of cases. It consists in describing crucial social conditions that constrain social policies, decisions and, in general, interactions, as well as examining the applicability of philosophical or social studies theories to these interactions

The book's second task is primarily normative. It consists in using the said conditions to assess the applicability of various social decision processes—from negotiation and bargaining to, most prominently, convention-settling processes—to a variety of contemporary issues, as well as to related problems in ethics and social and political philosophy, other branches of philosophy, and related social studies.

As evidenced in the table of contents, the essays' topics are wide-ranging. Some examine traditional and current positions in moral philosophy and related nonphilosophical studies as they apply to current policy-making problems. Others discuss traditional and current positions in other branches of philosophy and related nonphilosophical studies as they apply to social and personal problems raising widespread concern. But all the essays are united by their use of ecological concepts in addressing the problems. Many of these concern the interactions of business, technology, and society. Some, for example, are about the testing, production, and marketing of industrial and pharmacological products and the conflicts these create within firms and in the interactions between these business firms and society. Others are about environmental deterioration and its relation to energy and materials technologies, and safety criteria in environmental policy making.

But not all the problems addressed in the essays are prompted by technological developments or are policy-making problems. Some essays discuss problems concerning such matters as cross-cultural interactions among individuals; the ecological interactions—sometimes symbiotic relations—between humans and their habitats and how this affects culture and nature: and the social convention-settling processes involved in a variety of human activities—from reaching a certain balance with the natural environment to coming to accept new or foreign art objects as art or even good art.

As always, I accept responsibility for my mistakes; but there would have been many more had I not benefitted from many discussions with friends and colleagues for the past few years. Among these, I owe the greatest thanks to Paul R. Churchill and Nancy E. Snow for their encouragement and sound advice at crucial times in the book's development. Thanks go also to the many people at various conferences with whom I tested some of the ideas formulated in this book. Among them, I would especially like to acknowledge the attendees at the 1994 Annual Conference of the Society for Philosophy in the Contemporary World, who commented on an earlier version of essays 1 and 2, and those at the 1995 National Conference on Ethical Issues in Finance, who commented on an early version of essay 5.

For the comments on topics related to my project that they made, at different times during the past few years, I am also grateful to David Braybrooke, the late Lewis A. Dexter, María C. Lugones, Enrique E. Marí, Jon N. Moline, Suzanne Stern-Gillet, and Jack Weir. I also thank my colleagues at Central Connecticut State University with whom I engaged in discussions relevant to the book. Among them, I am especially grateful to Parker English for his helpful comments on an earlier draft of essay 3. For their extremely effective library support services, I also thank the staff at the Central Connecticut State University Elihu Burritt Library, especially Emily S. Chasse, Richard Churchill, Norma Chute, Marie A. Kascus. Barbara Sullivan Meagher, Faith A. Merriman, Joan G. Packer, Nicholas G. Tomaiuolo, June Sapia Welwood, John Cayer, and Tzou-Min Hsiung. I should also mention and thank the Connecticut State University for providing funds and a sabbatic leave in support of the research that led to this book and the Central Connecticut State University College of Arts and Sciences, for granting reassigned time to pursue the project.

What I owe my wife, Mary Kay Garrow, for her unfaltering encouragement, experienced policy-making comments, sound editorial suggestions, and even sharing the responsibility for writing essay 5, I cannot possibly repay. As for our daughters, Alejandra Emilia and Catalina Patricia, they have, as in the past, played an essential part in all this by simply being with us, which has filled my life with happiness, and by their joy of life, love of learning, and growing search for their place in the universe, which give concrete meaning to this book's aspirations.

A.P.I.

NOTES

1. Jess Stein et al., "ecology" entry no. 2 in *Random House Dictionary of the English Language* (New York: Random House, 1970), p. 452.

2. Lewis Anthony Dexter, "Intra-Agency Politics: Conflict and Contravention in Administrative Entities," *Journal of Theoretical Politics* 2 (1990): 151–72.

3. David Braybrooke, *Meeting Needs* (Princeton, N.J.: Princeton University Press, 1987).

4. Edith Cobb, *The Ecology of Imagination in Childhood* (New York: Columbia University Press, 1977).

5. Margaret Mead, "The Cross-Cultural Approach to the Study of Personality," in J. L. McCary, ed., *Psychology of Personality* (New York and London: Grove Press and Evergreen Books, 1959), p. 233.

PART 1

The Multiculturalism Issue and the Ambiguities of Culture*

A Fragmented Social World

In our fragmented twentieth century, diversity and growing cultural self-assertion elicit romantic enthusiasm among some and cautious skepticism—if not outright fear—from others. There are those who hail pluralism as a way out of ethnocentric parochialism and see in it a new freedom for long-oppressed ethnic groups. Others argue that pluralism is just another name for confusion and that this confusion is undermining the very foundations of Western culture.

Philosophy has not remained impervious to this controversy. Some traditionalists are determined to uphold the intellectual values embodied in the works of Plato, Aristotle, and other Western philosophical classics against non-Western influences. Some multiculturalists are equally determined to demonstrate that Western philosophical classics embody values that are detrimental to non-Westerners. Against both these separatist positions, some multiculturalists advocate integrationism among the various cultures and traditions involved. These differences of opinion lead to a variety of heated discussions and interactions that are part and parcel of what I will characterize as the *multiculturalism issue*: a cluster of sharp differences of opinions or conflicts of demands concerning cultural diversity and self-assertion, together with what people do to uphold their opinions or satisfy their demands.[1]

This issue poses a variety of questions, beginning with, What conception of culture is best suited for dealing with the multiculturalism issue? At what point in the inquiry prompted by the issue should we settle on a definition of multiculturalism? These are the questions I will address in this essay.

I will advance two theses. The conception of culture as a system of meanings is most likely to cover the various, more particular, conceptions of culture involved in the multiculturalism issue. However, the definition of culture can and should be left open until a sound approach for dealing with the

2

issue has been developed and, within this approach, a definition of culture adequate to the issue's nuances can be formulated.

In advancing these theses, the ambiguities of the term *philosophy* are likely to create confusion. There are at least four senses of this term: (1) *personal philosophy*, or a person's outlook on things, as in "My philosophy is not yet significantly formulated"; (2) *group philosophy*, or a group's predominant outlook on things, as in "Aymará philosophy includes the belief that time is circular"; (3) *philosophical theory*, or a generalized device for formulating, clarifying, and dealing with problems about reality, knowledge, reasoning, norms, and values, as in "Aristotle's and Confucius's philosophies are worth comparing with each other"; and (4) *a branch of inquiry*, or a critical activity about philosophies in the personal, group, and theory senses, as in "Philosophy is not an exact science." In this essay, I focus on philosophy in the branch of inquiry sense, as it applies in cross-cultural settings.[2]

AMBIGUITIES OF CULTURE

Ordinary Conceptions of Culture

Ordinary talk of culture is plagued with as much ambiguity as talk of philosophy. Such semantic diversity is a significant component of the multiculturalism issue.[3] Since distinguishing between the different senses in which the term culture is used might help reduce confusions that contribute unnecessary heat to the issue, I will briefly examine them next.

There is, for example, a *collective heritage* sense in which a culture is "developed in the historical experience of social groups" and, as social heritage, is passed on to succeeding generations intentionally.[4] That is, what is considered heritage is not just anything passed on, however indirectly and inadvertently this may happen. Instead, the group's members must generally endorse what, perhaps inadvertently, is passed on.

For example, suppose that, as some anthropologists argue, oedipal complexes are learned and shared widely in a number of cultures.[5] They still would not be part of these cultures in the collective heritage sense because they would not be generally endorsed by the cultures' members. In fact, they would be unrecognized, indirect consequences of the learning of other things.[6] Nor would just anything recognized by the cultures' members as an indirect consequence of the learning of other things be part of the culture in the collective heritage sense. For instance, kleptomaniac behavior is recognized across many societies as the indirect consequence of learning other things, but it is not generally endorsed in these societies. Hence, it is not part of their collective heritage.

Of course, things can change, and what was not part of a culture's collective heritage at one time can become part of that heritage at a later time. An example is the tango. As a dance, the tango appeared in the brothels of the Argentinian and Uruguayan shores of the River Plate in the 1860s. This origin hardly made it part of these societies' collective heritage. At least in Argentina, it was not generally accepted—certainly not by the upper classes—as part of the collective heritage until around the early 1910s, when tango became fashionable in France, thus earning, in the eyes of many of my compatriots, its certificate of cultural approval.[7]

Things can also change in the opposite direction, and what used to be part of a group's collective heritage may become a minor, uninfluential footnote buried in the group's past, so that, in effect, it ceases to be part of its collective heritage. An example of this is the way in which, with the wars of independence during the 1800s, the red and yellow colors that symbolized the Spanish crown ceased to be generally recognized as having any significance in Argentina, Uruguay, and various Central American countries. Instead, light blue and white took their place in the public imagination.

One might think that the collective heritage sense of culture is always honorific, but it is not. Some groups have a collective heritage that, at least in part, is nothing to be proud of. Examples of these groups are those whose culture included the practice of slavery.

In any case, this collective heritage sense, in which we talk of southeast Asian, sub-Saharan, Andean Precolombian, and Native American cultures, by contrast with Western cultures, is different from the *subgroup heritage* sense of culture in which we talk of, say, the jazz musicians' culture or, more accurately, subculture. But the difference is only one of scale, because all these groups can be said to have a historically developed, though perhaps more specifically focused, social heritage that they intentionally pass on to succeeding generations.

Other subgroups, however, can arguably be said to have a culture of their own in the sense that they have practices but no historically developed heritage they intentionally pass on to succeeding generations. The hippies of the 1960s and the beatniks in the United States, for example, can be said to have had a culture in this *practice* sense.

Cultures in the practice sense are not typically thought to prompt debate in the multiculturalism issue. By contrast, one crucial concern is respect for heritage. But this respect is both too wide and too narrow to cover all the concerns crucial to the multiculturalism issue. First, it is too wide because not just any group's heritage is at issue. That of jazz musicians is not. Second, it is too narrow because the culture of such groups as gays and lesbians is also involved in the multiculturalism issue, but not for reasons of heritage. Indeed these cultures are not, or not predominantly, thought to have a historically developed social heritage of their own that they intentionally pass on to suc-

ceeding generations. The reason why they are involved in the multiculturalism issue is that their way of life is being excluded and undermined, and
hence, gays and lesbians are not being treated with the fairness and respect to
which they are entitled. This suggests that, in fact, there is still another
operant sense of culture, the *way of life* sense, in which what is at issue is not
heritage but respect for and fairness to those who have a certain way of life.

One might object here that the fact that gays and lesbians are included in
current discussions of multiculturalism does not entail, and should not be
used as grounds for, the inclusion of gays and lesbians in the multiculturalism
issue. But this objection is one of the many ways of getting caught up in the
issue rather than soundly dealing with it. The multiculturalism issue is sometimes enhanced and its intractability increased by the fact that those involved
in various aspects of the debate fail to communicate with one another because
they are caught up in their own conceptions of culture and unable to think of
alternative ones. In addition, as exemplified by the latter objection, some
advocate one conception to the exclusion of the others. But this exacerbates
the issue and leaves little room for dealing soundly with it. Indeed, it places
every, or nearly every, group involved on a collision course with one another.
This is why, as current intercultural conflicts around the globe make plain,
there are many prudential and moral reasons against efforts to adopt and insist
on one of these narrow and inflexible definitions of culture. Hence, to the
extent that circumstances permit, attempts should be made at bridging the
conceptual gaps in a manner that is at least realistic; that is, at a minimum,
true to the diversity of conceptions and concerns involved in the issue.

ANTHROPOLOGICAL CONCEPTIONS OF CULTURE

In the hope of helping reason prevail in the multiculturalism issue by
attaining a more articulate overall conception of culture, one might look at
what anthropologists have to say about it. However, except for agreeing that
we all need culture, they are not entirely of one voice on the subject. Indeed,
in *Culture: A Critical Review of Concepts and Definitions*, Kroeber and Kluckhon
cite 164 definitions of culture![8] One prominent anthropologist, Roy G.
D'Andrade, describes the recent history of the term culture in anthropology,
linguistics, and psychology as follows:

> When I was a graduate student, one imagined people in a culture; ten years
> later culture was all in their heads. The thing went from something out
> there and very large to something that got placed inside. Culture became a
> branch of cognitive psychology; it was the content of cognitive psychology.
> We went from "let's try to look at behavior and describe it" to "let's try to

look at ideas." Now, how you were to took at ideas was a bit of a problem—
and some people said, "Well, look at language." . . . Before 1957 the definition of culture was primarily a behavioral one—culture was patterns of
behavior, actions, and customs. The same behavioral emphasis was there in
linguistics and psychology. The idea that cognition is where it's at struck all
three fields at the same time. . . . I think it was a nice replacement. But the
thing is now breaking . . . in psychology . . . in linguistics, and . . . in anthropology—and we each have different ideas about how it's breaking up.[9]

That the exclusive emphasis on cognition and language is breaking up,
as D'Andrade says, is evident from the fact that, currently, there are at least
three predominant conceptions of culture among social scientists. First, there
is culture as knowledge—a conception that identifies culture with the accu-
mulation of information, which need not be shared. Second, there is culture
as a cluster of norms and institutions—a view that identifies culture with
group structure. Third, there is culture as constructed reality—a position
that identifies culture with conceptual structures within which people con-
struct their world assuming it to be shared by other members of a group.[10]
Each one of these conceptions gives prominence to a characteristic function
that can also be found in language. Culture as knowledge emphasizes the
descriptive function. Culture as a cluster of norms and institutions emphasizes the directive function. Culture as constructed reality emphasizes the
emotive, affective, and creative functions of language.[11]

Given the variety of anthropological and ordinary conceptions of culture
involved in the multiculturalism issue, it is unwise to attempt to decide
which of these conceptions is best suited for soundly addressing conflicts
about cultural diversity. Each one might be, and seems to be, useful concerning a different aspect of multiculturalism. Indeed, echoing Geertz,
D'Andrade suggests that they all are aspects of an overall conception: that of
culture as a system of meanings.[12] This is a promising suggestion that
deserves further examination and provides reason in support of the essay's
first thesis: The conception of culture as a system of meanings is most likely
to cover the various, more particular conceptions of culture involved in the
multiculturalism issue.

However, as with the other conceptions of culture, embracing it at the
outset would be putting the cart before the horse. For what matters is developing a sound approach for dealing with the multiculturalism issue, not getting sidetracked into fine-tuning definitions of culture. Once the sought
approach is sufficiently developed, we will be in a position to investigate
what account(s) of culture are suited to this approach. In other words, the
accounts being discussed will stand or fall with whatever theory is eventually
formulated and survives critical scrutiny and related decision procedures.

One may wonder whether this can be done. The answer is affirmative.

One can—as I will—leave the definition of culture open and proceed to discuss the multiculturalism issue as it applies to a variety of groups that have some claim to be considered cultures, even though they would be excluded entirely from consideration based on inadequate definitions of culture. By adopting this approach, it will be possible to highlight the matters at stake in the issue, thus developing reasons for adopting a conception of culture that is sensitive to real concerns instead of arbitrarily begging questions that should not be begged. Hence, the reason for the essay's second thesis: The definition of culture can and should be left open until a sound approach for dealing with the issue has been developed and a definition of culture fitting this approach and adequate to the issue's nuances has been formulated.[13]

NOTES

*This essay is based on the first part of my "Dealing with Diversity: Cultural Fragmentation, Intercultural Conflicts, and Philosophy," presented at the 1994 Annual Conference of the Society for Philosophy in the Contemporary World on August 15, 1994, and included in Robert Paul Churchill, ed., *Crossing Cultural Boundaries* (Amherst, N.Y.: Humanity Books, forthcoming).

1. For a more detailed characterization of issues, see my "Issues and Issue-Overload: A Challenge to Moral Philosophy," in *Philosophy as Diplomacy: Essays in Ethics and Policy Making* (Amherst, N.Y.: Humanity Books, 1994), pp. 1–12.

2. I have discussed various senses of the term *philosophy* and how they relate to philosophy's aims, philosophers' motives, and philosophy's schools and traditions in *Through Time and Culture: Introductory Readings in Philosophy* (Englewood Cliffs, N.J.: Prentice Hall, 1974), pp. 1–8 and 452–61.

3. See, for example, Robert Fullinwider. "Multicultural Education," *Report from the Institute for Philosophy and Public Policy* 11, no. 3 (summer 1991): 12. For various characterizations of culture that emphasize historical aspects through heritage, see A. L. Kroeber and Clyde Kluckhon, *Culture: A Critical Review of Concepts and Definitions* (New York: Random House, 1963), pp. 89–94.

4. Melford E. Spiro, "Some Reflections on Cultural Determinism and Relativism with Special Reference to Emotion and Affection," in Richard A. Shweder and Robert A. LeVine, *Culture Theory: Essays on Mind, Self, and Emotion* (Cambridge and New York: Cambridge University Press, 1984), p. 323.

5. Melford E. Spiro, *Oedipus in the Trobriands: The Making of a Scientific Myth* (Chicago: University of Chicago Press, 1982).

6. Roy G. D'Andrade, "Cultural Meaning Systems," in Shweder and LeVine, op. cit., p. 114.

7. For a discussion of this and related points, see my *Through Time and Culture*, pp. 405–409. For a brief Spanish-language account of the historical development of tango, see José Gobello, "tango" entry in *Diccionario Lunfardo* (Bueños Aires: A. Peña Lillo, 1975), p. 202. For evidence of how fashionable tango became in France in the 1910s, see Sem, *Tangoville sur mer: Every body is doing it now!* (Paris: Success, 1913). For an English-language social history of tango, see Donald S. Castro, *The Argentine Tango as Social History* (Lewiston, N.Y.: E. Mellen Press, 1991).

8. Kroeber and Kluckhon, op. cit., p. 291.

9. D'Andrade, op. cit., p. 7.

10. Ibid., passim, especially pp. 115–16.

11. D'Andrade, loc. cit.

12. See, for example, Clifford Geertz, *The Interpretation of Cultures: Selected Essays* (New York: Basic Books, 1973), passim, and *Local Knowledge: Further Essays in Interpretive Anthropology* (New York: Basic Books, 1983), p. 182.

13. A parallel point can be made concerning the concept of nation, which is sometimes used instead of that of culture.

Dialogue

— I understand the distinctions you draw between different senses of the term *culture*. I can also understand why, at this stage in your discussion, you want to leave the definition of culture open. But I do not understand what you hope to accomplish with this approach. Isn't the definition of culture a mere matter of semantics?

— A matter of semantics, it is. A mere matter of semantics, it is not. For, at the very least, words have consequences and, in the case of the term culture, a consequence of people's using it with different meanings in their discussions of multiculturalism is confusion, when not conflict.

— Yet, why would people involved in the multiculturalism issue be moved by meaning distinctions? They may be too emotionally involved to pay attention to such minutiae. Or they may already know the distinctions but stick to their own meaning preferences.

— No doubt, some people do this. But not all of them do. For there are those who, rather than stubbornly sticking to their preferred meanings, are interested in doing away with the confusion and conflict that their using meanings differently than those used by others tends to cause. They would rather clarify the matter so as not to be talking at cross-purposes with those who hold opposite opinions. The distinctions I have drawn would certainly be of some help to these people.

— Perhaps, but this is a long way from helping address the multiculturalism issue. How much mileage do you expect to get out of your discussion on the meanings of *culture*?

— I only expect to achieve two things. The first, as I just indicated, is a modicum of clarification that would, however modestly, contribute to cooling off the discussions among some of those people who are involved in the multiculturalism issue. Second, by holding that the definition of culture should be left open for the time being, I expect both to avoid an unending and sterile semantic discussion at the outset and to begin to provide the a groundwork for the eventual formulation of a sound definition.

— That is not much.

— Right. This is why this first essay is the shortest in the book. I begin to address more substantial concerns in the second essay.

— I am curious to see how you do that. But you have not quite addressed a point I briefly made before. As you granted me, some people may already know the distinctions you draw but stick to their own preferred meanings. You have simply said that others do not do this. Right. But your answer does nothing by way of dealing with the issue created by the fact that, at least concerning meanings, the people I described play a different language game from yours. How would you settle matters between these games? Can you?

— Here, I can only respond with two points. First, these so-called games are not mere games because, as I said, words have consequences. The so-called games affect people's lives and, hence, are not mere games. Second, even if the way in which some people involved in the multiculturalism issue use the term *culture* conflicts—perhaps to the point of being currently incommensurable—with the way in which other people use the same term, this conflict can be pragmatically resolved through interactions among the people involved. This is where, as I shall argue and illustrate in the book's remainder, convention-settling processes are crucial.

— I was in search of a straightforward answer and, instead, I end up with a bunch of questions. What is a convention-settling process? How can it help resolve incommensurabilities between different linguistic uses? Can it help resolve more substantial conflicts? How? And how is the process of such resolution pragmatic?

— Good. I'm glad you asked these questions. I will address them, and others, throughout the remainder of the book. Let us begin to address these question with regard to the multiculturalism issue.

Dealing with Diversity: Cultural Fragmentation, Intercultural Conflicts, and Philosophy*

THE MULTICULTURALISM ISSUE, POLITICS, AND PHILOSOPHY

The multiculturalism issue would be well on the way to being resolved if it were primarily the result of ambiguities in the use of terms such as *culture*, and its derivatives, not the least of which is *multiculturalism* itself. But the issue goes much deeper than that, posing very substantial questions. In the previous essay, I dealt with the question, What conception of culture is best suited for dealing with the multiculturalism issue? The present essay seeks to answer, What political and philosophical position is more defensible in dealing with the issue: separatism, integrationism, or something else? Also, it is the burden of this essay to offer a satisfactory response to the question, What changes in the current conceptions and practice of philosophy are needed so that philosophy can offer a more positive influence in dealing with the multiculturalism issue?

I will pursue four theses. I will argue, first, that separatism is inadequate for dealing with the multiculturalism issue and incompatible with philosophy, and, second, that integrationism is too strong a position to help address the fact of cultural fragmentation and the multiculturalism issue. Third, consequently, I will maintain that the issue can better be addressed dialectically, that is, through meaningful cross-cultural dialogue and social interactions such as negotiation, arbitration, mediation, and consensus building. These arguments all point to a fourth and final thesis: In order to be helpful—rather than irrelevant—in dealing with the multiculturalism issue, philosophy can, and needs to, make room for meaningful crosscultural dialogue and interactions, to aim at humanly attainable comprehensiveness, and to interlock with the social sciences and the nonacademic world.

11

CULTURAL SEPARATISM, PHILOSOPHY, AND SOCIETY

Since the 1950s, cultural diversity has been dramatically communicated to the growing number of people all over the world who have access to a television set. The feeling of cultural diversity has been reinforced throughout the planet as a result of technology transfers, international business efforts, and policy developments that have made fax machines available to ordinary people from southeast Asia to the southern cone of South America (and led to the widespread use of the metric system in most countries). Yet, as the African philosopher K. C. Anyanwu, echoing many others, puts it:

> The whole world seems to have become united under the metric system as well as the system of "hook ups" and "plug ins," but the spiritual distance between nations seems to have increased enormously.[1]

Among the manifestations of this fragmentation, Anyanwu mentions the intransigence of nationalism. As for the factors that increase the fears and tensions concomitant with nationalism and the conflicts it creates, he correctly identifies the denial or lack of discussion of issues concerning cultural experience and values, human dignity and integration, and human coexistence and tolerance.

One way in which these issues are denied or explained away by misdescription (and fruitful discussion is subsequently avoided) is through *cultural separatism*. This position is sometimes described as the view that cultural communities must be represented only in their own voices and from their own point of view. Thus understood, separatism can have many, sometimes quite opposite, social uses. For example, in the early 1970s, an Argentinian friend of mine and I were leaving the United States and, while the plane was taking off, my friend started a conversation with a U.S. passenger. After the usual exchanges concerning national origin and reasons for visiting the country, the conversation turned to culture and politics, and my friend expressed some views sympathetic to the U.S. counterculture movement. The passenger's answer was curt and, in the above sense, separatist: "You talk about your country, and I'll talk about mine."

In this sense, the recent (and not so recent) practice and teaching of philosophy in the United States and other Western countries was significantly separatist. Exclusively Western authors, from Plato and Aristotle to Bentham, Nietzsche, Ryle, and Heidegger, were discussed and their problems and positions examined in accordance with the canons of some Western philosophical school—say, analytic, phenomenological, or pragmatic.

To be sure, comparative philosophy was practiced as a recognized area of study. But many an established philosopher might have taken issue with whether it was philosophy ordinarily understood or, rather, a higher-level and extraphilosophical area of inquiry. Indeed, as recently as 1991, Nalini Bhushan said of the comparative philosophy approach in relation to the predominant practice of philosophy in North America:

> Notice, however, that this is an alternative that in effect requires that we give up our basic philosophical practice in favor of a quite different kind of philosophical practice which takes the form of a comparative critique. . . . Philosophy as we do it, is unlikely to be able to maintain a genuine commitment to take on culturally diverse philosophies when it comes to the question of style . . . because our philosophical style is essential to our philosophical practice.[2]

I have no doubt that conflict between much current philosophical practice and comparative approaches is real in at least Europe, the United States, and Canada. Yet things are changing too fast to believe that such practice is—if it ever was—predominant, let alone basic or essential, as Bhushan suggests. Nonetheless, philosophical fragmentation is real and largely parallels the conceptual, sometimes politically charged fragmentation I previously described in ordinary as well as anthropological thought concerning culture. As the debate over cultural diversity in philosophy mirrors fragmentation in the larger society, examination of the controversy in philosophy helps illuminate the different positions that might be taken on the multiculturalism issue. Concerning philosophy, then, and in the midst of its current fragmentation, the question arises, Can there be a defensible place for separatism in philosophy today?

Upon considering the preceding discussion, one might argue that there is a narrower sense of separatism I have not considered. It is the view that the various cultural communities that were traditionally excluded must be represented in their own voices and from their own point of view. This position is often described as making room for non-Eurocentric voices and perspectives. In discussing this form of separatism, Robert Fullinwider distinguishes two versions—weak and strong—as they apply to the United States. Concerning the "official" American story, he says:

> The *weak separatist* might believe that . . . any "corrective" enrichments from excluded points of view are bound to be subverted into mere emendations and embellishments, indeed, subverted into buttresses for the story's main themes. For the present, previously excluded points of view need their own undiluted voices, their own uncompromised forums.
>
> A *strong separatist*, in contrast, would repudiate the very aspiration for a reconstructed common story of America and Americans. There is no common perspective to be found, there are only diverse independent perspectives.[3]

Fullinwider criticizes strong separatism on the grounds that it implies that "an Arawak-centered perspective is no better than a Eurocentric perspective or no part of a more comprehensive synthesis." The point, Fullinwider argues, is not that a Eurocentric perspective is better but that strong separatism embraces such a strong version of cultural relativism that it becomes impossible for members of any culture to criticize any other culture. Strong separatists cannot consistently criticize Eurocentric culture for excluding their own perspectives. Thus, on this view, "the label 'Eurocentric' ceases to be a charge, complaint, or criticism; it becomes merely a description."[4]

This criticism of strong separatism is incontestable. However, the question still remains: Why not be a weak cultural separatist? According to this position, ideally, it is possible, and would be best, to make room for dialogue and mutually corrective enrichments between different cultures. But it is not feasible in the current circumstances. For these efforts are likely to backfire by undermining the said nonseparatist, cross-cultural project to the advantage of predominant Western views.

Regarding philosophy, there is no doubt that both the currently predominant languages, concepts, methods, and conceptions, and the very practice and social organization of philosophy—through such things as its departments and programs, associations, journals, and referee networks—may, and sometimes do (some would say often) undermine nonseparatist, cross-cultural projects. There is also no doubt that such difficulties provide reasons not to take part in specific projects because they would lead in highly undesirable directions. However, this situation provides no good reason for doing this sort of thing as a rule; that is, the difficulties do not provide a defense of weak separatism.

In fact, there are good reasons against weak separatism in philosophy and, generally, in society. First, separatists distance themselves from efforts to find at least the common ground needed for meaningful dialogue and interaction between different cultural traditions. Because they avoid any attempt at facilitating meaningful dialogue—philosophical or otherwise—until circumstances change, they actually undermine efforts to initiate dialogue across cultural traditions so that circumstances improve. Thus, weak separatists trap themselves in a vicious circle, for circumstances will never change without such dialogue and the variety of social interactions that go with it and help us learn ways of avoiding useless conflict. In short, weak separatism, in philosophy or in society at large, undermines the attainment of its own ideals.

Second, separatism in philosophy, whether non-Western or Western, ignores the history of Western philosophy and the purposes that originally guided it. Thomas Auxter makes this case as follows:

> The origins of the Western philosophical tradition are themselves multicultural. . . . There is a line of continuity and development in Western

values that leads to a recognition of the value of cultural diversity. . . . The method of amplification outlined by Plato is based on an inclusive approach to the world's wisdom.[5]

One might add that not only the origins, but also the various stages of Western philosophy were cross-cultural. For example, Averroes's "interpretations of Aristotle in their Latin form influenced philosophers in the Christian West from the thirteenth to the seventeenth century."[6] This is not an isolated example. The Jewish scholastic philosopher Maimonides influenced Aquinas and Eckhart. Indian thought influenced Schopenhauer.

Third, even if one is skeptical about these claims, the fact still remains that the search for wisdom through dialogue and the use of reason without any limitations or restrictions are characteristic of philosophy since its beginnings. In their attempts to find principles of truth and right not merely accepted by convention, Socrates, Plato, and Aristotle criticized the ethnocentric myths and religion of their society. The dialogue and reliance on reason they sought, and philosophy as the search for wisdom seeks, cannot be suspended without seriously undermining—if not entirely suspending—the activity of philosophy itself. It is primarily here that weak separatism and philosophy as a branch of inquiry must part company. Of course, it may turn out that, once the inquiry has been carried on long enough, irresolvable cultural differences remain. But there will still be room for further philosophical inquiry concerning what should be done about those differences.

Finally, philosophy would stand to gain from engaged dialogue across cultures in at least three ways. First, such reflection would lead to greater understanding of the nature, scope, and purposes of philosophy, and attaining this understanding is an important aim of philosophy. Second, such reflection would help us better assess philosophy's present condition and future prospects. Third, this kind of reflection would help all those who engage in it develop more open and inquiring minds and engage in philosophically fruitful cross-cultural dialogue. Hence, there are good reasons for meaningful—that is, not merely a pretense of—cross-cultural dialogue in philosophy. Indeed, in this age of growing cross-cultural interactions, it is urgent that we all become more open-minded and engage in fuller cross-cultural dialogue. As history and current events make plain, cultural parochialism often creates devastating conflicts. All forms of separatism are at odds with this fact because, as a rule, they institute the conditions that lead to further conflict. Hence, the essay's first thesis follows: Separatism is inadequate for dealing with the multiculturalism issue and incompatible with philosophy.

These remarks, however, are only a beginning—and on this matter we are very much at the beginning—toward cultural harmony in philosophy, and in society. If one is to proceed, a further question must be addressed:

How can we proceed in a nonseparatist and cross-culturally sensitive and fruitful way? I will address this next.

THE INTEGRATIONIST RESPONSE

As a reaction to cultural separatism, some writers have argued for a form of pluralism that is sometimes called *cultural integrationism*. A notable advocate of this position is Diane Ravitch. Here are her views on the multiculturalism issue in the United States:

> The pluralists promote a broader interpretation of common American culture by recognizing first that there is a common culture, and second that it has been created by many different groups. At its most basic, our common culture is a civic culture, shaped by our Constitution, our commitment to democratic values, and our historical experience as a nation. In addition, our very heterogeneity . . . creates styles of expression that the rest of the world perceives as distinctively American. The cohesive element in the pluralistic approach is the clear acknowledgment that, whatever our differences, we are all human.[7]

Ravitch's alternative to separatism, however, is not very helpful in dealing with the multiculturalism issue. There are various reasons for this. First, the civic culture she finds to be common in the United States is, at best, a very limited component of what might count as the United States' collective heritage. To be sure, the United States is not a country founded on ethnicity or nationalism, but one founded on a constitutional structure that, in principle, makes room for a great diversity of peoples. But this is, at most, a political and legal component of the United States' heritage Ravitch seeks identifying, and it will hardly suffice to address the multiculturalism issue in the manner she proposes. For the nature and, indeed, existence of a thick network or well-structured mosaic of shared beliefs, norms, institutions, and practices supposed to constitute the entire heritage is what is at issue.

Second, Ravitch uses the phrase "our historical experience as a nation." This implies that, at some level, it is common. However, as any separatist will be quick to point out, no matter how we stretch it, the said historical experience is hardly the same for such groups as Native Americans, who were largely excluded from the predominant society; African Americans who were forced to take part in slavery; and other quite disparate participants such as Hispanics, Italian Americans, and Anglo Americans. No doubt, members of each of these groups were sometimes involved in the same historical events in various ways. But beyond participating in the same historical events, their actual role in and experience of these events was, and arguably remains, quite fragmented and conflictive.

Third, Ravitch's point about the distinctive styles of expression prompted by United States cultural heterogeneity is well made, but primarily aesthetic, and hence too limited to complete the identification she seeks of a full, thick collective heritage for the United States. Besides, as separatists will also point out, like jazz, such expressions are largely the style of particular groups and often express those groups' frustration with their lot in the United States.

Fourth, Ravitch's exclusive emphasis on culture as collective heritage is inadequate to deal with the issue and, as previously argued, trying to embrace the narrower cultural heritage conception at the outset would be putting the cart before the horse. Whether, given our definitional preferences, we like it or not, culture as a way of life is also crucial to the issue, and a realistic philosophical analysis of culture should be true to this fact and not get entangled in possible distinctions that distract from the actual multiculturalism issue and help worsen it rather than soundly address it. Indeed, Ravitch's exclusive emphasis on cultural heritage is a glaring example of the narrow positions criticized in essay 1. For it advances a too narrow and inflexible conception of culture to help address the multiculturalism issue. Instead, one should address the multiculturalism issue by fairly and squarely addressing all the conflicting concerns it involves, without getting sidetracked into fine-tuning abstract definitions of culture, let alone ruling out some candidates by dogmatic definitional fiat.

Altogether, these considerations point to a more crucial reason why Ravitch's position is not helpful for our purposes: For the resolution it seeks requires convergent, generalized agreement on a variety of matters of belief and value. That is, it requires a very substantial common ground in the sense of a thick network, or a well-structured, somewhat fixed, also thick mosaic of shared beliefs, norms, institutions, and practices, which, if existent or attained, would amount to a single national culture. But why believe such a convergence is likely? In fact, why believe it is necessary in order to address the issue in a sound manner?

One should adopt a much more parsimonious approach. After all, the only common ground needed is that which furthers meaningful cross-cultural dialogue and social interactions—that is, not a mere pretense of dialogue and reliance of reason—between those groups and individuals involved in the cross-cultural conflicts that prompt the issue. No less, but no more.[8] This is especially significant at the international level, where hopes of finding or developing a common culture are likely to add wood to the fire instead of helping resolve the issue. Indeed, as some of my acquaintances from Quebec attest, through desperate disillusionment, integrationism often implodes into separatism.

The preceding discussion provides reasons for accepting the essay's next two theses: Second, integrationism is too strong a position to help address the fact of cultural fragmentation and the multiculturalism issue: and, third, the issue can better be addressed dialectically, that is, through meaningful cross-cultural dialogue and other social interactions, such as negotiation, mediation, arbitration, and consensus building. Here, the question arises, What changes in the conceptions and practice of philosophy are needed in order for philosophy to deal constructively with the issue? I will address this in the ensuing discussion.

CROSS-CULTURAL DIALECTICS AND PHILOSOPHY

Various conceptions of philosophy can be found competing with one another today. Some philosophers conceive of philosophy as *a research field* left over from when the sciences went their independent ways. Others think of philosophy as *the study of large unsolvable problems*.[9] Still others describe philosophy as an art "that uses for its medium not stone or paint and canvas or sound, but argument."[10] There are also those who conceive of philosophy as *a dialogue about fundamental problems*. And there are those who think of philosophy as *the task of underlaborers*, with philosophers working with scientists to help them with conceptual difficulties so that they can solve problems in their respective sciences.[11]

From the standpoint of helping us deal with the multiculturalism issue, these conceptions of philosophical practice are too narrow and constrictive. Take the conception of philosophy as a research field. Suppose that, in some sense, philosophy is a research field just like physics and biology, only that it aims at gaining characteristically philosophical knowledge, say, about the mind-body problem. This goes no way toward helping to deal soundly with the multiculturalism issue. For it cuts philosophy off from the cross-cultural dialogue and social interactions, which, as previously indicated, are involved in dealing soundly with the issue. In order to help us deal with the issue, philosophy needs to be conceived in a manner that allows it to engage and, hence, become sensitive to, the various concerns and ideas actually involved in multiculturalism. For these same reasons, conceptions of philosophy as the study of large unsolvable problems, as an art, and as a dialogue about fundamental problems are also inadequate. They all fall far short of the mark.

The underlaborer conception of philosophy, however, might escape the criticisms just formulated, because it joins hands with nonphilosophers and, supposedly, helps them address a variety of practical and theoretical problems. Kai Nielsen describes this conception.

> The underlaborer's conception of philosophy goes well with a commitment to a type of analytic philosophy that has no systematic pretensions but is content instead to dispel conceptual confusions emerging out of everyday life and

science. We are to do this, it is often claimed, with the aid of sharp new analytical tools that philosophy as conceptual or logical analysis provides.[12]

He grants that "philosophers with a good analytic training have a developed capacity for drawing distinctions, spotting assumptions, digging out unclarities, seeing relationships between propositions and setting out arguments perspicuously." But, as Nielsen notes, the abilities he enumerated for philosophers with good analytic training are also displayed by "lawyers, classicists, economists and mathematicians with good training."[13] In other words, these abilities do not amount to unique philosophical contributions.

A critic could respond that the difference is not one of abilities but of the manner in and purposes for which the abilities are put to use. One could claim that philosophers use these abilities in a nondogmatic manner and in order, not simply to convince, but to establish which views pass the test of critical scrutiny. Yet, though this function is commendable, it is hardly distinctive of philosophy. Branches of inquiry other than philosophy—for example, economics and mathematics—also display these functions.

Still, in contrast to other branches of inquiry, philosophy aims at as much comprehensiveness as can be humanly attained, or at least at integrating the results and activities of these other branches of inquiry in an overall conception of our world. This is not to say that philosophy aims at providing a privileged view of the universe. Nor is it to say that every philosopher should become a generalist instead of focusing on specific areas such as the philosophy of biology or the philosophy of psychology. It is simply to point out that, as a collective enterprise, philosophy aims at formulating, clarifying, and dealing with problems that concern the nature of reality and mutual relations between beliefs, reasons, values, and norms.

Nielsen is skeptical of comprehensiveness in philosophy and remains largely silent about the philosophical aims just described. His position concerns more limited philosophical prospects:

> There may be a way that philosophy might transform itself in a way that would answer to our unschooled reflective hopes. It would involve (a) giving up all pretensions to autonomy and instead interlocking philosophy fully with the human sciences and (b) taking the resolution of the problems of human life to be very centrally a part of philosophy's reasons for being. . . .
>
> What I am advocating . . . is a holistic social theory which is at once a descriptive-explanatory social theory, an interpretive social theory and a normative critique. Departing radically from the philosophical tradition, it will be an empirical theory. Elements of the social sciences will be a very central part, although, in light of the importance of giving a narrative account of who we were, are, and might become, much of the social science utilized may be historiographical.[14]

I sympathize with Nielsen's suggestion and believe that, as far as it goes, it is sound doctrine. However, it does not go far enough. If the problems of human life are to become central to philosophy, the interlocking Nielsen suggests must involve the nonacademic world as well as the social sciences. Otherwise, philosophy will be unable to address problems—for example, those posed by the multiculturalism issue—thoroughly and realistically. Some concerns will likely fall through the conceptual and methodological cracks of the different languages of philosophy, other academic disciplines, politics, technology, and so on. Also, the general public will be unrepresented and their concerns might well be missed.

Some attempts to bring academics and nonacademics together into specific activities that involve philosophy have been carried out—for example, through hospitals' ethics committees in health care ethics, and through committee hearings and advisory offices in the legislative process. The results are instructive and point to problems likely to arise. They concern the process of selecting participants; the perceived incompatibilities between disciplinary loyalties, loyalties to particular traditions, or personal styles; academicians' fears of being co-opted by representatives of government, industry, or labor; their uneasiness about becoming entangled in conflicts between government, industry, or labor factions; and the unwillingness on the part of practitioners of this or that profession or trade to submit their practice to academic or public scrutiny.[15]

These conflicts should be expected. Indeed, they are likely to be greater when, as in the multiculturalism issue, conflicts between different cultural sectors are involved. Philosophy, after all, is not an apolitical activity, especially in the highly institutionalized environment in which it is pursued today. Squarely acknowledging and sincerely addressing such conflicts is central to the development of philosophy along the lines suggested here. This is even more crucial when the people involved have different cultural or social backgrounds.

Under these conditions, it is not sufficient—indeed it may not always be fully possible—to integrate the concepts, methods, and practices of different disciplines, or academic with nonacademic approaches to the problems. What is crucial—whether in order to attain further integration or to make sound decisions in its absence—is to establish a fruitful interplay between different—sometimes sharply divergent—cultures or social sectors.[16]

The interplay envisioned here involves not merely—sometimes not even primarily—dialogue, but also, as with married couples learning to be partners in a good relationship or as foreigners learning to live in a foreign culture, a variety of interactions that help them learn to live well together, even when a thorough integration of concepts, methods, or practices is not attained, or attainable. This completes the argument for the essay's fourth—

and last—thesis: In order to be helpful—rather than irrelevant—in dealing with the multiculturalism issue, philosophy can, and needs to, make room for meaningful cross-cultural dialogue and interactions, to aim at humanly attainable comprehensiveness, and to interlock with the social sciences and the nonacademic world.

To argue for these changes is not to say that basic research in philosophy and the current organization of such research must change. Nor is it to claim that everyone doing philosophy must start doing things differently. Rather, it is to argue that philosophy must undergo some changes—at least some extensions—as a collective enterprise. Indeed, the needed changes are not just organizational. They are significantly theoretical and include a conceptually integrated formulation, clarification, and evaluation of alternatives for dealing with cultural fragmentation and the multiculturalism issue. This must be a result of a collective, multioccupational activity—not the privileged subject of any one philosophy researcher, educational practitioner, or policymaker. Nor can it be the privileged subject of any one philosophical school, discipline, profession, or occupation.

The fact that these matters are not the privileged subject of any such sector provides additional support to the view that the previously indicated interplay is crucial and requires changes in the collective practice of philosophy. Without including all relevant constituencies, the problems are unlikely to be addressed in a manner that is sensitive to all concerns involved.

One might object here that the preceding discussion shows only that, in order not to become hopelessly irrelevant concerning cultural fragmentation and conflicts, academic disciplines need to join efforts with other disciplines and nonacademic sectors. But the discussion does not show that the public needs such academic efforts. If the problems are real, they will be addressed by ordinary people anyway. They will do so with or without the help of aloof academicians or busy policymakers.

In response to this criticism, one may say that there is good reason to believe that policymakers, however involved in the process, or the general public, however informed it may become about cross-cultural matters, will not have such an abundance of time or even sufficient wisdom to make all other contributions unnecessary. This is where academic—though, of course, not merely academic—efforts can help.

At least because of its longstanding experience with this type of situation, anthropology should play a role in the process. It has a lot to contribute to clarifying the very notion of culture, formulating the distinction between culture and race, understanding such things as kinship systems and ethnocentrism, and helping, through ethnographic studies, to develop a reliable wealth of cases. In this way, it would contribute to free inquiry from ethnocentric, naive, and romantic (however well-meaning) interpretations, such as the notion that Western thought emphasizes reason while non-Western thought is based on tradition.

History's role should also be central because this discipline has a lot to contribute to placing problems in the context of the times in which they are formulated. And this context is crucial not only for understanding the problems, but also for establishing what would be solutions to them. The philosophical inquiry here envisioned should proceed in this anthropologically and historically informed direction. Such an approach would help us deal more intelligently, and less parochially, with the multiculturalism issue.[17] It also offers us humans hope of finding a way to rise above our fragmented times.

NOTES

*This essay is based on the second and third parts of my "Dealing with Diversity: Cultural Fragmentation, Intercultural Conflicts, and Philosophy," presented at the 1994 Annual Conference of the Society for Philosophy in the Contemporary World on August 15, 1994 and included in Robert Paul Churchill, ed., *Crossing Cultural Boundaries* (Amherst, N.Y.: Humanity Books, forthcoming). My thanks go to all the conference participants for their comments, and, especially, to Professor Robert P. Churchill for his careful reading of and comments on the original version.

1. K. C. Anyanwu, "Cultural Philosophy as a Philosophy of Integration and Tolerance," *International Philosophical Quarterly* 25 (September 1987): 271.

2. Nalini Bhushan, "The Real Challenge of Cultural Diversity: Clarifying the Boundaries of Legitimate Philosophical Practice," *Teaching Philosophy* 14 (June 1991): 170–71.

3. Robert Fullinwider, "Multicultural Education," *Report from the Institute for Philosophy and Public Policy* 11, no. 3 (summer 1991): 13–14.

4. Ibid., p. 14.

5. Thomas Auxter, "Toward Multicultural Philosophy," *Teaching Philosophy* 14, no. 2 (June 1991): 191–92.

6. Julius Weinberg, *A Short History of Medieval Philosophy* (Princeton, N.J.: Princeton University Press, 1964), p. 139.

7. Diane Ravitch, "Multiculturalism Yes, Particularism No," *The Chronicle of Higher Education* (October 24, 1990), p. A44. See also Diane Ravitch, "A Culture in Common," *Educational Leadership* 49, no. 4 (December 1991): 8–11; "Diversity and Democracy: Multicultural Education in America," *American Educator: The Professional Journal of the American Federation of Teachers* 14, no. 1 (spring 1990): 16–20, 46–48; "Integration, Segregation, Pluralism," *American Scholar* 45, no. 2 (spring 1976): 206–17; *The Revisionists Revised: A Critique of the Radical Attack on the Schools* (New York: Basic Books, 1978).

8. I have extensively discussed the forms and nature of such dialogue and interactions in *Philosophy as Diplomacy* (Amherst, N.Y.: Humanity Books, 1994), passim. I have also discussed and provided examples and interdisciplinary contributions relevant to the nature and forms of meaningful dialogue and interactions in *Contemporary Moral Controversies in Technology* (New York and London: Oxford University Press, 1987) and *Contemporary Moral Controversies in Business* (New York and London: Oxford University Press, 1989).

9. Marjorie Grene, "Puzzled Notes on a Puzzling Profession," *Proceedings and Addresses of the American Philosophical Association*, vol. 61, supp. (Newark, Del.: APA, 1987), pp. 75–76.

10. Ibid., pp. 76–77.

11. Ibid., pp. 77–78.

12. Kai Nielsen, "Philosophy as Critical Theory," *Proceedings and Addresses of the American Philosophical Association*, vol. 61, supp. (Newark, Del.: APA, 1987), p. 95.

13. Ibid., pp. 96, 97.

14. Ibid., pp. 98–99.

15. See, for example, Arthur Caplan, "Can Applied Ethics Be Effective in Health Care, and Should It Strive to Be?" *Ethics* 93 (1983): 311–12. For whatever it is worth, my modest experience in academia—in philosophy, interdisciplinary studies, and the development of interdisciplinary course curricula and workshops—confirms the existence of the conflicts mentioned in the text. But it has also convinced me of the viability of multioccupational approaches and the need for negotiation and other social decision procedures to play a central role in them. For a more detailed discussion of this matter, see my *Philosophy as Diplomacy*, "Bridging Gaps in Babel: Ethics, Technology, and Policy-Making," pp. 181–93; "Like the Phoenix: Ethics, Policy-Making, and the U.S. Nuclear Energy Controversy," pp. 175–78; "A Delicate Balance: Reason, Social Interaction, Disruption, and Scope in Ethics and Policy-Making," pp. 106–108; and "Practical Equity: Dealing with the Varieties of Policy and Decision Problems," pp. 143–44.

16. In this regard, the conception of philosophy as diplomacy, which I have characterized and substantially discussed elsewhere, should be of help. In a nutshell, philosophy as diplomacy addresses the policy and decision problems posed by issues "in ways that are feasible, effective, and crucially sensitive both to the often unsettled and conflictive nature of the concerns that contribute to pose the problems and to the variety of open-ended social decision procedures that may help settle these concerns and deal with the problems through policies and decisions, and on the basis of reasons worked out in the process." For a more detailed discussion of this conception, see essay 6 in my *Philosophy as Diplomacy*, pp. 70–84. The quoted passage appears on pp. 75–76. Applications of this conception and discussions of how it contrasts with other philosophical approaches can be found throughout the book.

17. This concern has been expressed by anthropologists. See, for example, Richard J. Perry, "Why Do Multiculturalists Ignore Anthropologists?" *The Chronicle of Higher Education*, March 4, 1992, p. A52. For interdisciplinary discussions of culture, see Richard A. Shweder and Robert A. LeVine, *Culture Theory: Essays on Mind, Self, and Emotion* (Cambridge and New York: Cambridge University Press, 1994).

Dialogue

— I am puzzled by your laconic remark concerning Fullinwider's criticism of strong separatism. You merely say that his criticism is incontestable. Perhaps it is. But why?

— I thought it was obvious and did not think it necessary to comment any further after having quoted Fullinwider's criticism. I will tell you now. Strong separatism holds that the perspective of each culture is incommensurable with the perspective of any other culture. If so, then there are only different cultural perspectives, none of which can be critically assessed from any other cultural perspective. As a consequence, strong separatists, who are characteristically involved in the multiculturalism issue in an attempt at upholding the values of their own culture, have no grounds for criticizing any cultural perspective that may conflict with their own, and can only describe them as different.

— Yet, they can still prefer their own culture, can't they?

— Only in the sense that they can like their own culture and not like the others or like them less. But this is merely a psychological fact and, given the strong separatism position, not a ground for assessing those other cultures as worse or less preferable than the separatist's culture. On this position, there is at most a mere diversity of mutually incommensurable culturally relative likes and dislikes. The cultural gaps between them cannot be bridged in any manner whatsoever.

— Can't these likes and dislikes, however, constitute reasons for criticizing those cultures one dislikes?

— They may be motives that lead the separatist to criticize other cultures; that is, they may be *his* or *her reasons* for doing so. However, given that, according to strong separatism, these reasons are irretrievably and totally culturally dependent, none of them can be *a reason* in the sense of making a difference from a standpoint that is not merely peculiar to the separatist's culture.

— Why should this matter though?

24

— Because it follows that, from a strong separatist standpoint, the question, What culture is better among those compared and why? makes no sense. Nor does the question, Why is that culture, which as part of its traditions holds itself superior to others, including our own, wrong in holding this? A consistent strong separatist should reject both in favor of, at best, What culture do I like more among those compared? Here, liking one culture over another would be a strictly arbitrary and unchallengeable matter, just as whether one likes vanilla ice cream better than chocolate ice cream.

— And so what? Aren't expressions of cultural preferences arbitrary?

— I don't think they are. Cannibalizing foreigners or enslaving them is not simply a matter of taste. It is subject to critical scrutiny and assessment on the basis of reason. Besides, if a strong separatist believed expressions of cultural preferences thus are arbitrary, he or she should not care to judge what other cultures or ethnic groups do or think. Yet, strong separatists involved in the multiculturalism issue do care to judge these matters and do not believe they are simply expressing their own tastes. Hence, strong separatism in effect becomes self-refuting. It is like meaning to be taken seriously when stating, "No statement should be taken seriously."

— Even if you are right on this point, however, how can cross-cultural, though nonintegrationist, dialogue be meaningful as you suggested?

— What do you have in mind?

— You say the only common ground needed is that which furthers meaningful cross-cultural dialogue between individuals and groups involved in the conflicts that prompt the multiculturalism issue. How could such a dialogue be possible without the substantial generalized agreement of both beliefs and values that integrationism postulates?

— Dialogue can be meaningful even when there is little agreement between those engaged in dialogue. For, despite sharp disagreements, it can involve interactions that are fruitful by at least building familiarity and mutual adaptations among those engaged in the dialogue. This makes the dialogue meaningful despite the fact that it may lead to no immediate agreement on matters of content.

— But wouldn't it at least be better to attain a thick agreement rather than the thin common ground you suggest?

— There are reasons to think it would not be better. First, as I argued in the essay, especially at the international level, hopes of finding or developing such common ground are likely to add wood to the fire rather than helping resolve the multiculturalism issue. Second, even at the national level, such hopes may have the same effect. Some of my acquaintances from Quebec tell me that the history of this Canadian province in the 1980s and 1990s evidences the fact that integrationist approaches may, upon their recurrent failure, lead to desperate turns toward separatist stands. Third, even if such a common ground could be developed, given how long it would take to bring about such a generalized conversion toward shared views and values, and how pressing it is to resolve the multiculturalism issue within a relatively short time, focusing on developing a thick common ground would be a distraction likely to backfire into failure and discouragement. Fourth, and perhaps most important, integrationists who seek to develop such common ground, and think it best, do not celebrate diversity. Rather, they put up with diversity until, through such thick common ground, they get rid of it. But, beyond the thin common ground I think necessary to deal with the multiculturalism issue's conflicts, diversity in culture, as in biology, is, at least generally, both necessary and fruitful. For, as I said before, dialogue can involve interactions that build familiarity and mutual adaptations among its participants even if it may lead to no immediate agreement.

— Your response is too vague. How is such a feat supposed to happen?

— I will begin to explain this in the next essay. It focuses on mutual adaptations and the development of interaction networks (or, as I shall say, social and related ecologies) involving people who belong to different cultures. In later essays, I will further describe this ecological model and apply it to other areas of human activity. Please bear with me.

— OK. Go ahead; but I'm skeptical.

— That's fine. Just keep an open, however critical, mind.

Cross-Cultural Ecologies: The Expatriate Experience, the Multiculturalism Issue, and Philosophy*

AN AMBIGUOUS PARADISE

The expatriate experience, with its unstable mixture of melancholy and exhilaration, permeates the end of the twentieth century. The experience itself is not new. Indeed, it has been recorded or orally passed down through the centuries in the stories of people of Jewish or Gypsy descent. Yet the scope and magnitude of this experience in the twentieth century is quite unprecedented, at times not only exceeding all expectations but moving cross-cultural realities beyond the wildest speculations of fantasy and fiction.

It partly results from massive migrations occurring over decades, such as the influx of European and, later, other immigrants who, in the 1880s, began migrating to countries such as Argentina and the United States. It also results from sudden exile brought about by a variety of events. Religious, ethnic, or cultural persecution, for example, led Bosnians who were able to escape Serbian ethnic cleansing into exile in the United States or Western Europe in the 1990s. And, throughout the twentieth century, coups d'état, civil wars, and political repression, sometimes coupled with rapidly deteriorating national economies, have significantly contributed to the growing influx of immigrants from Eastern Europe and northern Africa into Germany; from the Caribbean into the United States; from some Spanish-speaking countries into other Spanish-speaking countries; and from every Spanish-speaking country into a range of other countries, including Australia, France, and, especially, the United States, where this influx has led to the rapid growth of the Hispanic diaspora.

Migrations such as these need not occur only between countries. As in the cases of Peruvian peasants migrating to Lima, Argentine peasants migrating to Bueños Aires, and Puerto Ricans migrating to the U.S. mainland, they can occur within countries as well. In the introduction to *Paradise Lost or Gained? The Literature of Hispanic Exile*, Fernando Alegría poignantly

27

characterizes the scope of the resulting expatriate experience: "Exile no longer has borders."[1]

This experience is part and parcel of the multiculturalism issue: a cluster of sharp differences of opinions or conflicts of demands concerning cultural diversity and self-assertion, together with what people do to uphold their opinions or satisfy their demands.[2] In the sometimes heated discussions that constitute it, there are those who hail cross-cultural experiences and interactions as ways out of ethnocentric parochialism and see in them a new freedom for long-oppressed ethnic groups. Others argue that such intermingling of cultures and the position that supports it, pluralism, are merely forms of confusion and that this confusion is undermining national economies—say, those of Germany and the United States—and, indeed, the very foundations of Western culture.

In this essay, I discuss some salient characteristics of the expatriate experience and ask, Can they help us to understand and deal with the cultural fragmentation and cross-cultural conflicts involved in the multiculturalism issue? How? Can they help to delineate philosophy's role in dealing with the fragmentation and conflicts? I put forth two theses. First, some salient characteristics of the expatriate experience are (1) ambiguity about one's evoked or hoped-for homeland; (2) a fluid transculturational balance—in other words, a changing equilibrium in the expatriate's interactions with and adaptations to his or her foreign environment; (3) a heightened awareness of local conventions; and (4) an increased reliance on convention-settling processes—that is, on social decision processes, from argument and negotiation to quiet resistance and outright confrontation—whose aim is to settle new conventions in the expatriate's encounters with foreign environments. The second thesis is that, given their analogy with some salient features of philosophy, the characteristics of the expatriate experience can also help to delineate philosophy's role in dealing with the fragmentation and conflicts involved in the multiculturalism issue.

In the process of defending these theses, I will formulate and provide some reasons for adopting two working hypotheses. One hypothesis is that the features characteristic of the expatriate experience can help us to understand and deal with the cultural fragmentation and cross-cultural conflicts involved in the multiculturalism issue. The other hypothesis is that these characteristics can help by indicating the concerns characteristically involved in the multiculturalism issue—for example, the concern with retaining one's sense of personal identity while attaining one's aims in a foreign environment. The characteristics also help by focusing on convention-settling processes that are used in dealing with the issue. Arguments for treating these working hypotheses as actual theses will be offered in essay 11.

This essay presents and discusses a variety of stories or cases, proceeding in accordance with two research methods used throughout this book. One is taxonomic and has been used in both philosophy and social process studies. It consists in merging description and argument. That is, it describes a number of cases or observations to exemplify categories, classifications, and theories and then uses the same (or similar) cases or observations as evidence for the utility or accuracy of the categories, classifications, and theories.

One might think this approach involves a vicious circle. However, it does not. It involves no circle, or, if it does, it is not vicious. For, first, the same cases or observations used to exemplify the' categories, classifications, and theories are obvious (because trivial) evidence for the utility or accuracy of the categories, classifications, and theories regarding those cases and observations. To be sure, this shows no general utility or accuracy. But, second, it is here that similar cases or observations come into play. They provide evidence for the general utility or accuracy of the categories, classifications, and theories.

This taxonomic method is currently being used in social science where, for example, Lewis Anthony Dexter has described and used it to study contravention and conflict in administrative entities.[3] It is also currently being used in philosophy among philosophers who attempt to connect conceptual analysis or phenomenological studies with the results of scientific research, the experiences of the general public, and the aims of ethics and sociopolitical philosophy.[4]

The other ecological method was introduced by Edith Cobb in *The Ecology of Imagination in Childhood*.[5] Cobb used this method, which is based on the ecological model of interchanges and mutual adaptations and interconnections between organisms and their environment, to analyze children's development of cosmic sense. As Margaret Mead commented, however, the method has a wide scope of applicability and "makes it possible to compare the most diverse cultural developments."[6] Indeed, it may help us to understand not only, at the individual level, the mutual adaptations the expatriate experience involves between individual expatriates and the foreign cultures in which they live, but also, at the societal level, the adaptations developed between groups whose respective cultures differ. This latter understanding can, for example, be attained by relying on analogies between the mutual adaptations between these groups and the manner in which different species develop symbiotic relations.[7]

Some of the cases and observations I will discuss in applying these methods will be extralinguistic, but I will begin by focusing on actual language use in cross-cultural settings.

FROM TRAVEL-LANGUAGE TO LANGUAGE-TRAVEL

In his book *The Intellectual Life*, Phillip Gilbert Hamerton tells this story:

> When I lived in Scotland, three languages were spoken in my house all day long, and a housemaid came to us from the Lowlands who spoke nothing but Lowland Scotch. She used to ask what was the French for this thing or that, and then what was the Gaelic for it. Having been answered, she invariably asked the further question which of the three words, French, Gaelic, or English, was the right word. She remained, to the last, entirely incapable of conceiving how all three could be right.[8]

The woman had traveled and had been exposed to foreign languages—indeed, to three of them every day during a significant period of time. Yet, her presuppositions about them prevented her from knowing them well, if at all. This is an instance of a general fact about cultural crossings: When going abroad and being surrounded by a foreign language and culture, one misses more about them, the narrower one's preconceptions—and one cannot help having some—in approaching them.

I am not referring simply to the difficulties in learning a foreign language that are prompted by the mismatch between its syntactical conventions and those of one's native tongue—for example, that of learning to say, as in English, that one arrives *in* a city, not, as in a literal translation from Spanish, that one arrives *to* a city. Nor am I talking about superficial semantic difficulties such as that of remembering that the English terms for the Spanish *gato*, *mesa*, and *estudiante*, are cat, table, and student. I am talking about the variety of emotive meanings and functions of a language's expressions one might miss, even if one has learned its syntactical conventions and basic vocabulary quite well, because of one's preconceptions in approaching the language.

These preconceptions rarely clash with ordinary purposes that are unobjectionable and highly motivating for learning a foreign language—say, that of touring a foreign land or engaging in commerce abroad. However, here and there, they lead to misunderstandings, perhaps accompanied by social discomfort and even tension that can upset a vacation or undermine a business deal.

Consider, for example, the purpose of touring a foreign land, camera in hand, enjoying its picturesque features while, at the same time, keeping one's habits and native conceptions entirely unaffected by the experience. It can be fun, and many do just that. The linguistic knowledge it generally requires is merely that needed to get meals and a place to sleep, to understand directions, to engage in polite conversation and various forms of entertainment, and, depending on the circumstances, to take a taxicab to go through, say,

Lima, or to hire a guide and rent a donkey to reach Machu Picchu. In other words, all one needs to know is *travel-language*.

As stated, this is an unobjectionable purpose and may motivate one to learn a great deal of a foreign language. Yet, it is too narrow to help open one's eyes to the many contextual nuances in the use one makes of what one learned. Besides, it will not keep travelers entirely out of trouble. Suppose, for example, that with this purpose in mind, one goes to Argentina and, at a home whose members follow traditional gaucho customs, is offered a maté, the gaucho green tea. Suppose that, after drinking it, one returns the container saying, quite politely and, linguistically speaking, correctly, "gracias." As a result of this apparently impeccable behavior, those who offered the maté will be convinced that he or she did not like it. For, in that context, besides being a polite expression of thankfulness for the treat, "gracias" means that one does not want any more maté and, people normally do not say this until after they have had a few. In short, mere travel-language does not always travel well.

Of course one could, as many a wise traveler does—or as many an immigrant or immigrant descendant living in Argentina has done—learn along the way, perhaps asking others why (fortunately if one disliked maté, which, after all, is an acquired taste) no more matés were offered after the first. However, in order to learn about this convention and be able to perform the appropriate speech act if suitable circumstances arise again (and, in Argentina, or in an Argentine's home abroad, they probably will), one must curb the habit of saying "gracias" whenever one would have said "thank you" in English or "grazie" in Italian or, for that matter, "gracias" in practically any other Spanish-speaking environment. Instead, one may develop the habit of saying that the maté "está bueno" (is good), or something of the sort.

This amounts to changing, however minimally, those of one's habits and native conceptions that concern the manner and circumstances in which gratitude is politely expressed. As soon as one begins to do this, the purpose mentioned above has been dropped and one acts in accordance with a more flexible purpose. This new purpose makes room for the possibility that the experience of traveling through a foreign land and communicating with its people in their language may not leave one's habits and native conceptions unaffected. Such *linguistic convention-travel* is a second way of traveling with a foreign language and the starting point of *language-travel*.[9]

Of course, linguistic convention-travel does not have the consequence that, upon returning to one's home country, one will, for example, use "thank you" or "grazie" differently. Yet, barring forgetfulness and lack of reflection, one's ability to use polite forms of thankfulness well will not any more be entirely patterned on one's native usage or conceptions. One is now a bit less provincial than one used to be because, however modestly, through a short linguistic trip, one has moved beyond simply leaving one's country while, culturally, staying home the entire time. One has—to be sure, quite modestly and minimally—traveled through cultures.

This second way of traveling with a foreign language is certainly needed in order to move beyond the bounds of our native and initially unavoidable provincialisms. But it carries us only so far. It may make us happy tourists or successful businesspeople. Yet, it characteristically keeps cross-cultural interactions at a largely superficial, detached level.

Dissatisfied with these limitations, one may approach foreign language learning not just with the purpose of merely touring foreign countries or of simply engaging in commerce with foreign peoples, but with the purpose of understanding these countries and peoples. This, however, requires that one use the language significantly from *within* its cultural context in dealing with the world. Here, using language merely to translate or successfully communicate what one would say in one's native tongue, or merely to do the proper foreign act, becomes secondary. By contrast, learning the full range of nuances of what the language emotively means and does becomes central. This is language-travel as *culture-travel*.[10]

In my experience, a good test of whether one has significantly reached this level is whether one has the ability to make jokes and have fights in the foreign language. It is then, when timing is central or the pressure is high, that translation and its companion, reflection, have little or no opportunity to function, and one handles the language one learned very much like one drives a car or steers a boat on a river, reacting to, rather than reflecting about, the immediate traffic or the sudden changes in the river's current.

Reaching this level of linguistic ability takes tedious training and, most of all, being immersed in a culture for an extended period of time—very much like an anthropologist or a sociologist engaging in participant observation.[11] Such cultural immersion may be worth pursuing for its own sake and, sometimes, leads to some good. For example, those who engage in such language travel may develop a greater appreciation for not just others' but their own native language and culture and a greater awareness of the varieties of provincialism. However, there are dangers. In the process, their understanding of their own culture and language becomes, if not distant, more contingent. Indeed, provided a sufficiently thorough cultural immersion for a sufficiently long period of time, their understanding of their own culture and language may degenerate into estrangement and the weakening, if not loss, of their sense of cultural identity. This effect may be temporary or may become permanent, leading to the lonely and painful experiences of a culturally severed life, paradigmatically exemplified in the expatriate experience, to whose examination I will turn next.

CONSTRUCTING OTHERS

The use of words has consequences, and sometimes the use of the same word in different linguistic contexts has quite disparate consequences. Take the Spanish word *gringo*, which is said to derive from the Spanish word *griego*, whose English translation is "Greek." In the United States, Mexico, and some areas of the Caribbean, gringo is an exclusively disparaging term, typically used among Hispanics to refer to foreigners, especially to U.S. nationals of British and Northern and Eastern European descent. In Argentina, however, the word *gringo* also means foreigner, but it can be used in either an endearing, a disparaging, or a mixed way to refer to foreigners or descendants of foreigners, especially those whose paternal last name is not Spanish.

A friend can say endearingly during a soccer game, "That was a great pass, gringo!" An example of its disparaging use can be found in the following story: A mestizo guide from Cerro Colorado, a small town in the mountains of the Argentine province of Córdoba, once was telling me that the Italian farmers in the area, upon buying their land and without any consultation with the locals, had enforced their newly acquired property rights by setting up fences around their farms to prevent free-ranging goats—including the guide's own goats—from destroying their crops. He stopped for a second and commented, "What would you expect from them? They are gringos." The implication was that gringos characteristically pursue their own interest without paying attention to the way in which their actions upset local traditions concerning land use.

As for mixed uses, one of the attendants at a party in Lobos, a town in the pampas of the Argentine province of Bueños Aires, once said upon hearing me play a gaucho drum: "Look at this gringo! He can play as well as or better than us!" The implication of such praise was that someone perceived as a foreigner had played well, and this was surprising because gringos were inept when it came to playing Argentine folkloric music.

This latter case is instructive concerning the manner in which Argentines tend to identify gringos. A lower- or middle-class Argentine descendant of Italians and Spaniards whose father's last name is Italian and whose mother's last name is Spanish—say, as in my case, Iannone Díaz—typically grows up believing that he or she is a Caucasian of Italian and Spanish descent who happened to be born in Argentina. The same will happen if the father's last name is Spanish while the mother's last name is Italian—say, Díaz Iannone. Yet, except for some upper-class families and some families of purely Spanish descent, Argentines tend to use only their paternal last names. Here, people will typically apply the term *gringo* only to the first case above.

Whatever the identification criteria, few if any social groups fail to use some of them. In doing so, they engage in an *ethnification process* that affects the sense of identity of children growing up and, indeed, of foreigners living among, or in contact with, these groups. After all, foreigners, just like children, have to learn many things from scratch. Suppose, for example, that the previously described Argentine whose first last name is Italian moves to the United States. Because of this last name, Hispanics, as well as non-Hispanics, in the United States would be quite unlikely to apply the term *gringo* to this person. Nonetheless, as, soon as be or she enters the country, the ethnification process begins. Given the history and the role of the legal system in the United States, the ethnification process is based on racial classifications reflected in legal classifications. The alien is required to fill out a form indicating his or her ethnicity in a manner that mixes race and culture. One of the choices is Caucasian. Another is Hispanic. The form explains that the box for Hispanic includes all those individuals from countries where the predominant language (as in Argentina) is Spanish. Caucasians are something else.

This is a construction, and one the Argentine previously described may have difficulty understanding. Though racially a Caucasian, he or she is legally characterized as Hispanic, not Caucasian. The initial response may amount to the unintended, or sometimes intended, failure to fill out the form as required. It may also involve a range of emotions: from bafflement, through amusement, to indignation. These may be enhanced, often in mutually conflictive ways, with an additional realization. In the United States, though an alien, this person is not perceived as a gringo or gringa but as a member of a varied group that includes, among many others, precisely the individuals who would use these terms to disparage him or her in an Argentine context.

This may not simply prompt conflicting emotions but also unsettle the person's sense of identity. After all, as an Argentine gringo or gringa, this person belonged to a group that, however disparaged, had acquired significant economic and political power in society. By contrast, as a Hispanic in the United States, the person belongs to a disadvantaged minority. The variety of intellectual and practical responses to this change in cultural constructs surrounding the newcomer centrally constitutes the expatriate experience, which I will examine next.

MODELING THE EXPATRIATE EXPERIENCE

Scope of the Expatriate Experience

The expatriate experience is as varied as the number of expatriates multiplied by the number of their distinct circumstances. Yet some common stages and salient features can be found within this variety. Fernando Alegría's account of his expatriation's first stage suggests some of these features.

When I left my homeland on September 23, 1973, there was no time to go out on the balcony and look at the white mountain range, the dovelike image of the Virgin of San Cristóbal Hill, the dusty treetops of the Parque Forestal; to sigh and silently wonder what fate held in store for me. One had to leave quickly, quietly, without looking back, wrapped in a scarf, if possible without showing one's face.

So I did what everybody does: I told my story, using the minor keys and matter of fact tone of a man who, having survived an earthquake, sips his drink, neither complaining nor protesting very much, just talking. The first months of numbness, pain and depression passed. I realized that a new phase of exile was beginning, that from now on there would be other periods, all different, each with its own anxieties, all shattering and over-whelming, and that I would be changing too, passing from one crisis to the next until I reached the moment of truth, unique and definitive—the day in which I would either stop being an exile and return home, or unavoid-ably, with sadness and resignation, become an immigrant.[12]

This passage encapsules three features commonly found in the initial stage of an expatriate's experience: numbness, pain, and depression. It also makes plain that the experience of exile need not exhaust the expatriate experience. For the experience of exile can be conceived, as Alegría conceives of it, as characteristically including the consciousness of exile, and this, in its turn, as crucially involving the intention to return home.

Though quite adequate for certain purposes, focusing strictly on exile is too narrow an approach for this essay's aims. For these concern not only cross-cultural interactions involving foreigners who intend to go back home, but also cross-cultural interactions involving cultural minority members who do not intend to go back home. Accordingly, I talk about the expatriate experi-ence in the wider sense that also covers, first, the experience of all immi-grants who do not intend to return home. These include those who, as Ale-gría puts it, have *become* immigrants. They also include those who, like Gyp-sies, have no homeland and, in many cases, no foreign home to go back to.

Second, I talk about the expatriate experience in the wider sense that also covers the experience of members of groups—say, members of the Jewish community in the United States or Argentina, African Americans, Native Americans, or Puerto Ricans in the United States, and peasant mestizos in Argentina and other American countries—who even when born in a given country are treated as foreign or marginal to the predominant culture. That is, I talk about the expatriate experience in the wider sense that also covers what sometimes is called "internal exile."[13] I will examine the experience's stages by appeal to cross-cultural studies of personality development.[14]

Varieties of Numbness, Pain, and Nostalgia

At least among exiles and immigrants, the ecology of expatriation begins with numbness, nostalgia, and sometimes also pain, even if accompanied by excitement. Numbness and pain can, though they need not, be relatively unreflective reactions to cultural shock brought about by relatively sudden changes in one's cultural environment. These changes are varied, including changes in the language predominantly used (for example, from Spanish to English); changes in the customary ways in which people relate to each other in public places (for example, how seriously they take the question "How are you?" or what noise level they consider to be loud); changes in the attention generally paid to the legal system; or the compound effect of all these and other similar changes. In contrast to numbness and pain, nostalgia, which can begin with unreflective mistakes such as expecting a type of food to be ready for dinner or a dear place to be just around the corner, develops along somewhat more reflective lines. Often—though by no means always—nostalgia is felt for the remembered homeland to which expatriates want to return.

In contrast to foreign expatriates, internal expatriates, depending on the degree of adaptation reached by the ethnic group in which they are born, may initially grow up accustomed to being treated like cultural foreigners, such as gringos in Argentina or Puerto Ricans in the United States. This may contribute to the absence of numbness and pain at least during the first years of life, before the individuals develop the ability to reflect. Later, however, numbness and pain may appear in response to their realization that they have been allotted undesirable social places through various constructs they find alien to their conceptions of themselves.

A further, more reflective development may appear with the natural desire to leave an environment where they are treated in such a manner, which, also quite naturally, may lead to reflection about their origins in the search for finding a place where they belong. Thus they may, and people do, imagine homelands. This notion may concern, as is the case among some African Americans, a homeland they have never visited. Or, as is the case among some members of the United States or Argentine Jewish communities, it may concern Israel, a place Jews now rule and which, before 1945 and for many centuries, they always identified as home in their collective memories.[15]

The Experience of Going Back Home

Many expatriates can, and some do, go back to the place they call home. The effects are often disquieting and, indeed, can be self-shattering. This can readily happen to internal exiles who have fantasized about their homeland for many years. For example, an Italian descendant born in Argentina, whose grandfather was from the eastern-central Italian province of Abruzzi-Molise and whose grandmother was from the southern Italian province of Potenza, may, upon "returning home" to Italy, have great difficulty coming to terms with the deep animosities he finds among a number of people from the provinces he had grown up to imagine as a harmonious and welcoming whole. The same thing may happen to an African American who goes home to an imagined Africa, only to experience great alienation upon being asked to what tribe he or she belongs and being scorned because he or she knows not what to answer. Whoopi Goldberg pointed to these disparities quite clearly and humorously when, during an Academy Awards ceremony televised in 1994, she said: "I've been in Africa and there is one thing I can assure you: I'm an American!"[16]

Going back home, however, need not be dramatic in order for it to affect one's sense of identity. Small, ordinary things may have the same effect. A Spanish speaker may go back home and find that, though still able to speak his or her native Spanish dialect—say, Argentine—he or she cannot understand it as readily as before because of having spent much time talking with speakers of some other—say Mexican or Venezuelan—Spanish dialect. Or this person may have developed a foreign accent and be told so by friends or family members. In any case, realizing that this has happened and that one's native language may not be as natural and as constitutive of one's own identity as one had assumed sometimes has a destabilizing effect.

Such destabilization is compounded by the fact that the longer the time one spends abroad, the less confirmable one's memories of one's homeland become. One's bed is not where it used to be any longer. One's cherished places have changed. Perhaps they have disappeared under the bulldozers and cranes of development. People have changed, perhaps becoming less trusting as a result of long years of oppression and civil wars. The list is long and it shows the impossibility of returning to much of the homeland of one's memory. This realization tends to have a significant effect on many expatriates' sense of identity. For this sense is highly dependent on memory, and the expatriate experience of going back home tends to collapse memories and fantasies together in the neverland of the unconfirmable. In this way, expatriates begin to find themselves living in a cultural and personal limbo.

Rejecting One's Homeland

The experience of living in a cultural and personal limbo leads some expatriates to engage in desperate adaptation efforts. They may turn their backs on their countries of origin, perhaps even on their families. And their attempts may lead to a variety of personality developments. Some may experience a buildup of internal conflicts with different degrees of tension and a variety of effects on their abilities to live full lives. Others may pragmatically live with rejection until forgetfulness and time allow them to get used to the idea that home, as they had imagined it, does not exist. Still others may reject their homeland, or dreamt homeland, without any upsetting consequences.

In any case, a significant number of expatriates—my impression is many, but by no means all—may eventually need to develop further adaptations to overcome the undesirable state in which their lack of fit with their homeland puts them. For example, they may need to accept the fact that, no matter how hard they try, they are still Argentines, or Norwegians, while also recognizing that they no longer fit in their homelands as well as they used to. They may need to acknowledge that they will never fit perfectly in their foreign environment, and they may need to identify areas of social life—say, marriage or business—in which they can function reasonably well in their foreign environment. At any rate, this process of seeking further adaptations is double-edged and varied, and involves much that is positive and much that is negative about the expatriate experience. Furthermore, it is not a one-shot or simple deal but typically takes place through a variety of transformations or cultural adaptations between individuals, between individuals and groups, and between the groups themselves.

Imagining Home, Transforming Oneself

The transformations characteristic of the expatriate experience concern the expatriate's past, present, and future and often manifest themselves in the expatriate's attitudes about his or her homeland, evidencing various forms of ambivalence. For example, María A. Salgado has substantially documented the fact that women poets of the Cuban diaspora "are aware that they have ceased to be part of Cuban history and that their poems, written in a foreign culture, are no longer part of Cuban literature."[17] Their ambivalence about Cuba is torn: Their portrayals of their past are idyllic, but they treat this past as destroyed and express deep alienation from the present and from themselves.[18]

By contrast, the Argentine writer Julio Cortázar, responding to the fact that his books were banned in his own country, displays a pragmatic ambivalence about the experience and about his homeland. He describes the experience as "not for me an altogether negative trauma," and talks of turning "the negative value of exile into a positive value."[19]

In both cases, the expatriates have significantly adapted to the realities of the foreign culture, thus achieving a somewhat fluid transculturational balance that includes their ambivalence about their homeland. This ambivalence, however, is not peculiar to cases in which expatriates accept the realities of expatriation, including the fact that it changes them. For example, the Puerto Rican writer Alfredo Villanueva Collado observes, "If one does, like I have, refuse to metamorphose, one must nevertheless choose some kind of official identity."[20]

Such an identity, however, is proposed by the reluctant expatriate as a way of settling conventions for presenting oneself in everyday life in the foreign culture. The success of this proposal is not a one-way street but instead involves give-and-take and mutual adaptations. Indeed, it cannot succeed without significant cross-cultural interpretation and interaction—a process that, like being married, changes people in many respects and not always in a manner they can easily articulate. What used to bother them—from people answering their questions with an "aha" instead of "yes" to people talking to them loudly upon hearing their foreign accent—may not bother the expatriates any longer. What once did not bother them—from an inefficient phone system to ethnic jokes—may bother them now. But even if they have not changed much or have not changed in these ways, at the very least, the process makes them more adept performers. And this is a disposition one cannot develop without a heightened awareness of local conventions or at least an increased ability to rely on the convention-settling processes involved in establishing new relations with people such as new roommates, new neighbors, or new coworkers.[21]

This is not to deny that rejection of the foreign culture can occur. Expatriates can have difficulty with, resist, or openly refuse to go along with a variety of cross-cultural adaptations. Some may refuse to adapt to new driving practices. Others may have difficulty adapting to personal relations entered into with different assumptions about love and loyalty. Still others may resist adapting to the foreign culture's decorum patterns whereby their homeland's ordinary level of talking is considered loud. Some—among whom I have found a number of writers whose intellectual life was closely tied with their native language—will simply refuse to learn the foreign culture's language.

Whatever the rejections, however, individuals must get on with their lives while in the foreign culture. As a result, a variety of adaptations need to, and do, occur. Indeed, they often serve as correctives for nonnegotiable cultural rejections. Further, they typically involve many personal changes, not the least of which is a heightened awareness of local conventions and an increased ability to rely on the convention-settling processes likely to help in the absence or ignorance of established conventions.

A WORD ABOUT CONVENTION-SETTLING PROCESSES

The question may arise whether convention-settling processes are, or are governed by, higher-level conventions. The answer is this: They may, but need not always be. Let us consider an example. Suppose that a foreign student has recently arrived in the United States from Argentina and has made a few acquaintances at a university campus. Also suppose that, while going to class, the student runs into one of these acquaintances and is asked "How are you?" The Argentine will act in accordance with his or her own country's convention and begin to explain how he or she is in a somewhat detailed manner. This type of answer may make the U.S. acquaintance somewhat impatient because the U.S. conventions governing those types of encounters call for quick answers such as "I'm fine. And you?" In addition, the acquaintance's impatience may be perceived by the Argentine as disinterest or rudeness.

Now, given that the newcomer may meet this acquaintance again, and will probably meet other acquaintances who will also behave in accordance with local—not Argentine—conventions, the convention mismatch needs correction. At the very least, the Argentine needs to learn the local relevant conventions. There are, of course, existing conventions that might be helpful in the given situation. For example, the Argentine might let others—say, compatriots more experienced in the ways of the local population, or a roommate belonging to the local culture—know of his or her bewilderment about the exchange, and they might explain the local conventions. But this does not settle the matter, for the newcomer might feel uncomfortable with the local conventions involved. To have learned them is not to have learned to live with them.

At this point, other existing conventions—say, those governing dialogue —might help the newcomer learn to live with local greeting conventions. But dialogue with compatriots might reinforce the newcomer's dissatisfaction with the local greeting conventions, while dialogue with locals might lead to further discomfort, if not conflict. After all, the locals may not understand or feel at ease realizing the newcomer's belief that such perfunctory greetings are rude and uncaring. What, in fact, is needed is the settling of conventions for dealing with each other's mismatches.

At this point, sheer interaction and its result, further and deeper acquaintance, may lead to familiarity and a new understanding of the behavioral mismatches concerning greetings. Such interaction may be governed by some additional conventions—say, those concerning generosity or kindness—but are not exhausted by them. The resulting familiarity, ease, and conventions for dealing with each other's mismatches—for example, that of not paying excessive attention to minutiae, or coping with sources of minor irritation—are partly a consequence of the frequent interactions themselves, not merely of the rules that govern them. In other words, though additional—and not necessarily higher-level—conventions may be involved, unconventional processes may also be, and often are, at work in settling new conventions for dealing with the initial convention (and related behavioral) mismatches.

This completes this essay's reasons in support of its first thesis. Some salient characteristics of the expatriate experience are (1) ambiguity about one's evoked or hoped-for homeland; (2) a fluid transculturational balance—in other words, a changing equilibrium in the expatriate's interactions with and adaptations to his or her foreign environment; (3) a heightened awareness of local conventions; and (4) an increased reliance on convention-settling processes—that is, social decision processes, from argument and negotiation to quiet resistance and outright confrontation, whose aim is to settle new conventions in the expatriate's encounters with foreign environments.

These characteristics of the expatriate experience invariably have an epistemological aspect—cross-cultural interpretations—and a practical aspect—interactions aimed at resolving conflicts or bridging communication gaps with others. But epistemological and practical aspects of this same type are at the center of the conflicts involved in the multiculturalism issue. They are part and parcel of the sharp differences of opinion or conflicts of demands concerning cultural diversity and self-assertion, which, as indicated at the outset, characterize the multiculturalism issue.

This interconnection provides reasons—which I will further detail in essay 11—for formulating two hypotheses. One is that the features characteristic of the expatriate experience can help us to understand and deal with the cultural fragmentation—which may range from linguistic and conventional mismatches to rampant segregation—and cross-cultural conflicts involved in the multiculturalism issue. The other hypothesis is that these features can help by focusing on the concerns characteristically involved in the multiculturalism issue—such as the concern with retaining one's sense of personal identity while attaining one's aims in a foreign environment. These characteristics also help by focusing on convention-settling processes, such as those previously mentioned—argumentation, negotiation, quiet resistance, and various forms of outright confrontation—which are used in dealing with the issue. With these hypotheses in mind, I will next turn to the role that philosophy can have in dealing with these issues.

PHILOSOPHY AS A FORM OF INTELLECTUAL EXPATRIATION

What is philosophy's role in dealing with the multiculturalism issue? When addressing this question, the ambiguities of the term *philosophy* are likely to create confusion. As discussed in the preface, there are at least four senses of this term. In addressing the above question, I focus on philosophy in the sense of a branch of inquiry and draw on some features of those who pursue it in a sustained manner.[22]

I have argued in essay 2 that if philosophy is to have any significant role in resolving problems of human life such as those posed by the multiculturalism issue, it needs to interlock fully with the human sciences and the nonacademic world. Otherwise, philosophy will be unable to address problems thoroughly and realistically. Some concerns will likely fall through the conceptual and methodological cracks of the languages of philosophy, academic disciplines, politics, technology, and so on. The general public will be unrepresented and their concerns might well be missed.[23]

A significant problem here, however, is how to interlock with the human sciences and the general public, which, in effect, is a public fragmented along cultural lines. The expatriate experience can help to clarify the place and role of philosophers and philosophy in this process. Let us see why.

Traditionally, philosophers, from Heraclitus, Socrates, and Mo-tzu to Ibn-Khaldun, Marx, and Gandhi, have approached their times and cultures with a critical, distant eye. This is no accident and it holds true even of philosophers who (as it is sometimes said of Aristotle, Aquinas, and Hegel) had the purpose of providing, or whose positions amounted to, a justification of their times, cultures, faiths, churches, or states. It is no accident because a critical, hence detached, approach is crucial to philosophy and the wonder with which it is supposed to begin. As Arthur E. Murphy put it:

> The philosophic mind is, in the first place, an inquiring mind. It is right and illuminating to say, with Aristotle, that philosophy begins in wonder. But not just any kind of wonder. The marvelous, the occult, the merely freakish and "colossal," are objects of wonder for the popular mind and arouse an interest that illustrated weeklies, motion pictures, sideshows and various forms of commercialized superstition exist to satisfy. The wonder that generates a philosophical inquiry is of a different sort—it is elicited not by the marvelous but by the familiar, by the things that everybody "knows" and takes for granted.[24]

This wonder about the familiar that everyone takes for granted turns philosophers into strangers in their times and cultures. They are a bit marginal, partly because they critically approach what these times and cultures assume to be known by everybody.

As evidenced by this essay's discussion of the expatriate experience, this marginal place and critical role of philosophers is quite analogous to the marginal place and critical role of many expatriates. For expatriates characteristically wonder at what everyone takes for granted. It is, for example, a common experience for recent expatriates who have come to the United States from Europe, Latin America, or Asia to wonder why, as previously discussed, upon running into U.S. acquaintances on the street or at a mall or campus and being asked "How are you?" and, accordingly, beginning to answer the question, the acquaintances often show a sort of restlessness, as if they had not meant to learn how the expatriate was after all. Of course, in the United States, casually asking "How are you?" in the said circumstances is largely equivalent to saying "Hi." Everybody knows this, takes it for granted, and, accordingly, gives a short answer such as "Fine. And you?" For recent expatriates, however, this is a source of wonder and, often, of criticism. They get the initial feeling that their acquaintances—perhaps all persons in the United States—simply do not care about others and cannot understand why they take the trouble to pretend to care by saying, "How are you?"

Granted, this is an example of a superficial misunderstanding that a little bit of time and listening can easily resolve. However, the example does exemplify the type of critical wonder that gets the inquiry started and begins to create feelings of estrangement. Further, the analogy goes deeper. Just as with expatriates, philosophers often experience a mixture of melancholy and exhilaration in doing philosophy; they develop adaptations to worlds they find, at least in some respects, foreign or unfamiliar enough to deserve examination; and, given their concern with the unfamiliar, they rely significantly on convention-settling processes—from proposing new social arrangements and ways of thinking about the world to suggesting and practicing ways of changing it. That is, just as individual expatriates, as well as entire cultures, make use of these processes in building a place for themselves in a foreign world, so too philosophy involves the use of the same kind of processes insofar as it aims at bringing about conceptual and practical changes in the world in which we live.

These features provide a certain commonality with those involved in cross-cultural conflicts caused by cross-cultural fragmentation and interactions. Hence, they provide a way in which philosophy can find a common ground that can help it to interlock with nonphilosophical activities. It also can help philosophers to communicate and engage in meaningful dialogue and convention-settling interactions with nonphilosophers involved in the multiculturalism issue. These are reasons for the essay's second and last thesis: given their analogy with some salient features of philosophy, the characteristics of the expatriate experience can also help to delineate philosophy's role in dealing with the fragmentation and conflicts involved in the multiculturalism issue.

The role here envisioned for philosophy cannot be discussed in detail within the constraints of this essay; but it will be further clarified in the book's remainder. Suffice it to say that it should proceed in accordance with the conception of philosophy as diplomacy, which I have developed elsewhere.[25] Let us briefly clarify this notion, beginning with the term *diplomacy*. By *diplomacy* I do not simply mean, as cynics would have it, the activity of doing and saying the nastiest things in the nicest ways. I mean, quite broadly, the activity of dealing with states, nations, and other social groups—such as the business sector, the financial community, the U.S. urban and rural communities, and the various ethnic groups in the United States—and even individuals, so that ill does not prevail.[26]

By *philosophy as diplomacy*, I mean a branch of inquiry aimed at dealing with a variety of problems and issues—such as cross-cultural problems and issues—in ways that are feasible, effective, and crucially sensitive both to the often unsettled and conflictive nature of the concerns that contribute to pose the problems and issues and to the variety of open-ended social decision procedures and processes that may help settle these concerns. These procedures and processes may deal with the problems and issues in such ways as negotiation, arbitration, mediation, and a variety of interactions that build familiarity, if not a common ground, of attitudes and reasons between the individuals and groups involved.[27] Among them are those involved in the convention-settling processes previously discussed.

An example of philosophy as diplomacy in action was already provided in essay 2, when suggesting a middle ground between the excesses of separatism and integrationism. As I then argued, this middle ground is a sounder alternative for dealing with the multiculturalism issue because it does not close in to dialogue (as separatism does), nor does it require convergent, generalized agreement on matters of belief and value (as integrationism does). Instead, the only common ground this more parsimonious alternative requires is that which furthers meaningful cross-cultural dialogue and social interactions among the groups and individuals involved in the intercultural conflicts that prompt the issue.

This emphasis on a parsimonious approach through open-ended dialogue and social interactions evidences two already indicated crucial features of philosophy as diplomacy. The first feature is that philosophy as diplomacy deals with problems in ways that are sensitive to the often unsettled and conflictive nature of the concerns that pose the problems and issues. The second feature is that philosophy as diplomacy is also sensitive to the variety of social decision procedures and processes involved in dealing with social problems and issues such as those concerning cross-cultural interactions.

To say that philosophy as diplomacy aims at avoiding ill might lead one to think that philosophy as diplomacy is primarily a consequentialist notion —that is, that it takes the justifiability of policies and decisions to depend only on the value of their consequences. But it does not. For ill may consist in the violation of a right or failure to act in accordance with a duty or with principles of justice. And these are deontological considerations—that is, they take the justifiability of policies and decisions to depend only on their accordance with rights, duties, or principles of justice.[28]

Nor is the conception of philosophy as diplomacy primarily deontological. It does not rule out the possibility that in certain cases—arguably in cases of widespread community deterioration—the social consequences are so catastrophic as to take precedence. And this need not be so because other deontological considerations take precedence. For the situations envisioned approach state of nature situations, and in any such situation, it is at least questionable whether deontological considerations such as rights or the obligations correlated with them carry much, if any, weight.[29] At any rate, the relative weight of these various considerations is at issue in such situations, and its determination needs to be worked out through the policy-making and convention-settling processes.

PHILOSOPHY AND ECOLOGIES

The preceding discussion suggests that philosophy's general role in dealing with cross-cultural and other social problems and issues would be in accordance with one of the methods described at this essay's outset: that introduced by Edith Cobb in her seminal *Ecology of Imagination in Childhood*. As stated there, Cobb used this method, which is based on the ecological model of interchanges and mutual adaptations and interconnections between organisms and their environment, to analyze children's development of cosmic sense. As Margaret Mead commented, the method has a wide scope of applicability and makes it possible to compare the most diverse cultural developments in a manner sensitive to the dynamics and constraints of cross-cultural ecologies.

Given the crucial role the term *ecology* plays in the present book, a more definite characterization than that provided by Cobb is in order. This characterization will be further developed throughout the book, but an initial version can be provided here. By an *ecology*, I mean a relatively settled network of interactive components. Following Cobb's and Mead's suggestions, the present book uses the term *ecology* in social, and not merely biological, contexts. Hence, the components of a *social ecology* can be not only, or primarily, a biota's components, but also, and crucially, beliefs, attitudes, practices, customs, institutions, policies, and social groups.

To say, as I have, that an ecology is relatively settled is to say that it is not merely episodic—a fleeting concatenation of components. This, of course, does not rule out the possibility that, on occasion, an ecology may be highly destabilized and approaching its downfall. Also, to say, as I have, that an ecology is interactive points to tensions among its components; but it is not to say that each component necessarily interacts with every other one. Just as with food chains, each component of an ecology interacts with some other component(s) but not necessarily, or often, with every one of them.

Philosophy's role concerning social and related ecologies need not be that of providing understanding alone. It can and sometimes, perhaps often, should also involve helping us to mold social ecologies in ways that make room for mutually supportive developments of the cultures and persons interacting in given social problems and issues, say, in the multiculturalism issue. For example, it should have such a corrective role whenever given social ecologies are to the serious disadvantage of some of those affected by involving parasitic or predatory relations between individuals or groups.

This corrective role is crucial given the curiosity of humans and their urge to reshape their environment. For if cultural purposes and social aims are not expanded to make room for the significant satisfaction of this urge among those involved in the issue, only conflict will remain. Indeed, as Edith Cobb puts it, the urge "will be displaced onto the ingenuity of delinquency and crime."[30] The cooperative, convention-settling focus of the approach here suggested should help to avoid these consequences and enable us to deal more intelligently and less parochially with the multiculturalism issue.[31] It also offers us humans hope of forging a home or, at the very least, a more welcoming social environment out of our conflictive times.

NOTES

*This essay has greatly benefited from discussions with the participants at the Philosophy and Cultural Diversity Conference of the Society for Philosophy in the Contemporary World, in Estes Park, Colorado, August 15–21, 1994. My thanks go to all of them. My very special thanks go also to Professors Nancy E. Snow and Parker English for their detailed and very helpful comments on earlier drafts of this essay. An earlier—much shorter—version of this essay, bearing the same title, appeared in Nancy E. Snow, ed., *In the Company of Others: Perspectives on Community, Family, and Culture* (Lanham, Md.: Rowman & Littlefield Publishers, 1996), pp. 191–208.

1. Fernando Alegría, "Introduction," to Fernando Alegría and Jorge Ruffinelli, eds., *Paradise Lost or Gained? The Literature of Hispanic Exile* (Houston: Arte Público Press, 1990), p. 15. For an indication of the extent to which migration is turning societies in every continent into multicultural societies in which significant numbers of individuals acquire transcultural self-images, see Scott Heller, "Worldwide 'Diaspora' of Peoples Poses New Challenges for Scholars," *The Chronicle of Higher Education*, June 3, 1992, pp. A7–9. Some significant recent studies concerning particular regions are Yossi Shain, *The Frontier of Loyalty: Political Exiles in the Age of the Nation-State* (Middletown, Conn.: Wesleyan University Press, 1989), and *Governments in Exile in Contemporary World Politics* (New York: Routledge, 1991); Richard G. Fox, *Recapturing Anthropology: Working in the Present* (Santa Fe, N. Mex.: School of American Research, 1991); Gay Wiletz, *Binding Cultures: Black Women Writers in Africa and the Diaspora* (Bloomington: Indiana University Press, 1992); Constance R. Sutton and Elsa M. Chaney, *Caribbean Life in New York City: Sociocultural Dimensions* (New York: Center for Migration Studies, 1992); Karla F. C. Holloway, *Moorings and Metaphors: Figures of Culture and Gender in Black Women's Literature* (Brunswick, N.J.: Rutgers University Press, 1992); Nina Glick Schiller, Linda Basch, and Cristina Szanton, *Towards a Transnational Perspective on Migration: Race, Class, Ethnicity, and Nationalism Reconsidered* (New York: New York Academy of Sciences, 1992); Hamid Naficy, *The Making of Exile Cultures: Iranian Television in Los Angeles* (Min-

neapolis: University of Minnesota Press, 1993); Ron Kelley and Jonathan Friedlander, *Irangeles: Iranians in Los Angeles* (Berkeley: University of California Press, 1993). For previous, general migration studies, see Kenneth C. W. Kammeyer, *Population Studies* (Chicago: Rand McNally, 1975), especially section 3, pp. 169–237.

2. For a more detailed characterization of issues, see my "Issues and Issue-Overload: A Challenge to Moral Philosophy," in *Philosophy as Diplomacy: Essays in Ethics and Policy Making* (Amherst, N.Y.: Humanity Books, 1994), pp. 1–12.

3. Lewis Anthony Dexter, "Intra-Agency Politics: Conflict and Contravention in Administrative Entities," *Journal of Theoretical Politics* 2 (1990): 151–72.

4. A relatively recent work significantly influenced by the analytic tradition, which exemplifies this approach, is David Braybrooke, *Meeting Needs* (Princeton, N.J.: Princeton University Press, 1987).

5. Edith Cobb, *The Ecology of Imagination in Childhood* (New York: Columbia University Press, 1977).

6. Margaret Mead, "The Cross-Cultural Approach to the Study of Personality," in J. L. McCary, ed., *Psychology of Personality* (New York and London: Grove Press and Evergreen Books, 1959), p. 233. Mead had seen Cobb's unpublished manuscript for this book and cited it as such in her 1959 article.

7. The literature abounds with studies of symbiosis as well as its applications to non-biological areas. For studies in biology, see, for example, A. E. Douglas, *Symbiotic Interactions* (Oxford and New York: Oxford University Press, 1994), and D. C. Smith and A. E. Douglas, *The Biology of Symbiosis* (London and Baltimore, Md.: E. Arnold, 1987). For studies in other areas, see, for example, David Cowart, *Literary Symbiosis: The Reconfigured Text in Twentieth Century Writing* (Athens: University of Georgia Press, 1993); Gary R. Edgerton, *Film and the Arts in Symbiosis: A Resource Guide* (Westport, Conn.: Greenwood Press, 1988); Margaret S. Mahler, Fred Pine, and Ann I. Bergman, *The Psychological Birth of the Human Infant: Symbiosis and Individuation* (New York: Basic Books, 1975); Kish Kurokawa, *The Architecture of Symbiosis* (New York: Rizzoli, 1988); and Hazel B. Pierce, *A Literary Symbiosis: Science Fiction/Fantasy/Mystery* (Westport, Conn.: Greenwood Press, 1983).

8. Phillip Gilbert Hamerton, *The Intellectual Life* (Boston: Roberts Brothers, 1875), p. 120.

9. Here I am conceiving of actual languages in their fullness, in which they are not exhausted by syntax and descriptive meanings but have an emotive and performative side, that of speech acts, whose functions depend on such conventions. See John R. Searle, *Speech Acts* (New York and London: Cambridge University Press, 1974), passim.

10. The notion of language-travel as culture-travel has a family resemblance to that of world-traveling in María Lugones, "Playfulness, 'World'-Travelling, and Loving Perception," *Hypatia* 1, no. 2 (summer 1987): 3–19. The term *culture* in culture-travel, however, is at least narrower than a world in Lugones's sense. For a culture cannot, while a world in Lugones's sense can, be an idiosyncratic construction of a society. Cultures are public. Also, the term *culture* in culture-travel refers to actual cultures, which arguably are ontologically unproblematic. World in Lugones's sense is used to describe experience, even if the content of this experience turns out to be ontologically problematic.

11. For an example of participant observation and its problems, see William F. Whyte, *Street Corner Society*, rev. ed. (Chicago: University of Chicago Press, 1955), pp. 299–307. As for the time needed to reach this level, it varies with the languages involved and the level of education the expatriate had at expatriation time. For a discussion of these matters concerning Spanish-speaking expatriates in the United States, see B. Bower, "Spanish Survives Bilingual Challenge," *Science News* 146 (September 3, 1994): 148.

12. Alegría, op. cit., p. 11. This is an excerpt from "Literature in Exile," *Review* 30 (September 1981): 10–23.

13. See, for example, Lucas Lackner, *Internal Exile: Poems & Illustrations* (Santa Barbara, Calif.: Santa Barbara Press, 1983); and Association of American Publishers, Inc., *Soviet Writers and Journalists in Labor Camp or Internal Exile* (New York: The Association, 1983).

14. See Mead, op. cit., pp. 223–24, 233; Cobb, op. cit., passim. In this regard, symbiotic relations between different species may serve as a submodel for this purpose. For studies of symbiosis and its applications to nonbiological areas, see note 7. For a philosophical attempt at integrating philosophical and ecological thinking, see Murray Bookchin, *The Ecology of Freedom: The Emergence and Dissolution of Hierarchy* (Montreal and New York: Black Rose Books, 1991); *From Urbanization to Cities: Toward a New Politics of Citizenship* (London: Cassell, 1995); *The Philosophy of Social Ecology: Essays in Dialectical Naturalism* (Montreal and New York: Black Rose Books, 1990): *Toward an Ecological Society* (Montreal and New York: Black Rose Books, 1980); and *Urbanization without Cities: The Rise and Decline of Citizenship* (Montreal and New York: Black Rose Books, 1992).

15. The existence of a homeland is not a requirement. Among some Gypsies, for example, a homeland may concern a place that does not exist and, arguably, is not supposed to have existed at least in any sedentary sense. For Gypsies were originally a nomadic, Caucasoid people who migrated from India and, throughout the centuries, settled for extended periods of time in Asia, Europe, and, more recently, in the American continent. That at least some Gypsies look—but do not look back—for a homeland is confirmed by the fact that, in the late 1960s, Gypsy representatives approached the Argentine government asking that Argentina allow them to establish a Gypsy homeland in the largely empty Atlantic shores of Patagonia. Their homeland is supposed to be, if anywhere, in the future. I saw the news of this request announced in Bueños Aires newspapers about 1966.

16. I witnessed Whoopi Goldberg's making these remarks during the March 21, 1994, Academy Awards ceremony, which she hosted, on the ABC network. In this regard, see also Eddy L. Harris, *Native Stranger: A Black American's Journey into the Heart of Africa* (New York: Simon & Schuster, 1992).

17. María A. Salgado, "Women Poets of the Cuban Diaspora: Exile and the Self," in Alegría and Ruffinelli, op. cit., p. 232.

18. Ibid., p. 230.

19. Quoted in Alegría and Ruffinelli, op. cit., pp. 9–10.

20. Quoted in Teresa Justicia, "Exile as 'Permanent Pain,' Alfredo Villanueva Collado's 'En el Imperio de la Papa Frita,'" in Alegría and Ruffinelli, op. cit., p. 183. See also Silvio Torre -Saillant, ed., *Hispanic Immigrant Writers* (New York: Ollantay Press, 1989), p. 40.

21. See, for example, Erving Goffman, *The Presentation of Self in Everyday Life* (Garden City, N.Y.: Doubleday, 1959), p. 251.

22. I have discussed various senses of the term *philosophy* and how they relate to philosophy's aims, philosophers' motives, and philosophy's schools and traditions in my *Through Time and Culture: Introductory Readings in Philosophy* (Englewood Cliffs, N.J.: Prentice Hall, 1974), pp. 1–8 and 452–61.

23. Besides essay 2 in the present book, I argued for this point in "Dealing with Diversity: Cultural Fragmentation, Intercultural Conflicts, and Philosophy," an essay presented at the Philosophy and Cultural Diversity Conference of the Society for Philosophy in the Contemporary World, Estes Park, Colorado, August 15–21, 1994. In this regard, the conception of philosophy as diplomacy, which I have characterized and substantially discussed in essay 6 of *Philosophy as Diplomacy*, should be of help. In a nutshell, philosophy as diplomacy addresses the policy and decision problems posed by issues "in ways that are feasible, effective, and cru-

cially sensitive both to the often unsettled and conflictive nature of the concerns that contribute to pose the problems and to the variety of open-ended social decision procedures that may help settle these concerns and deal with the problems through policies and decisions, and on the basis of reasons worked out in the process" (pp. 75–76). Applications of this conception and discussions of how it contrasts with other philosophical approaches can be found throughout the book.

24. Arthur E. Murphy, "The Philosophic Mind and the Contemporary World," *Reason and the Common Good* (Englewood Cliffs, N.J.: Prentice Hall, 1963), p. 368.

25. A. Pablo Iannone, *Philosophy as Diplomacy*; see note 23.

26. Iannone, op. cit., pp. 71–73.

27. Ibid., p. 73.

28. Ibid., p. 74.

29. Ibid., pp. 73–84.

30. Cobb, op. cit., p. 211.

31. Anthropologists have expressed the concern that the multiculturalism issue should be addressed in an anthropologically informed manner. See, for example, Richard J. Perry, "Why Do Multiculturalists Ignore Anthropologists?" *The Chronicle of Higher Education*, March 4, 1992, p. A52. For interdisciplinary discussions of culture, see Richard A. Shweder and Robert A. LeVine, *Culture Theory: Essays on Mind, Self, and Emotion* (Cambridge and New York: Cambridge University Press, 1994).

Dialogue

— How trenchant is the ecological model you propose? I know that by the term *ecology*, you mean a relatively settled network of interactive components. But this is vague. For one, What does it take for these components to form a network?

— All I mean is that the components are interrelated in a variety of ways. A biological example is that of food chains. Some components—say, some predators—affect other components (their prey) directly while they affect still other components (their prey's prey) only indirectly, as a consequence of their affecting the former. But this is not to say that all components affect all other components directly or indirectly. The networks may include subnetworks of components that hardly affect each other, just as the ecology specific to Tierra del Fuego in southern Argentina and Chile hardly affects the ecology specific to Taos, New Mexico, in the southwestern United States.

— This is fine and dandy for biological cases. But what about the social cases to which you apply the term *ecology*?

— The general concept of an ecology is the same. What changes, at least by extension to other types of components—such as beliefs, attitudes, practices, and customs—is its scope and, given that social and psychological components are now involved, the manners in which the components interact. The interactions need not be merely automatic any longer.

— This, however, raises questions about the specific manners in which they interact and how automatic and nonautomatic interactions affect or should affect each other.

— Of course. This is precisely one of the topics I will be discussing throughout the book. I will provide extended examples of social and related ecologies and, in addition, in essay 11, I will provide a conceptualization of such ecologies as a type of feedback system.

— This sounds like the systems approach.

— It may sound like it, but it is not. For, first, the ecologies I discuss are open-ended. Second, even if there can be open-ended systems, the ecologies

I discuss need not be, and often appear not to be, systems. They involve too much randomness and, as I said previously, too many mutually disjointed components to be systems in any useful sense of this term.

— OK. But your remarks lead me to ask many more questions. Does the model amount to anything beyond readiness to recognize multiple interrelated factors and some occasion to bring about mutual adaptation among them? Can the mutual adaptation sometimes be left to automatic processes, or does it always require intervention? Should the intervention always, or ever, aim at restoring automatic processes of mutual adaptation? Are some of these adaptations better than others? If so, does this imply a standard that is not itself subject to modification in the course of mutual adaptation?

— These are all relevant questions, and I will be addressing them in the book's remainder. Here I can only briefly indicate some of my answers. First, the model certainly goes beyond helping recognize multiple interrelated factors and occasions to act concerning mutual adaptations. It also helps explain what kinds of interrelations take place and what kinds of actions can and may be taken about them. Second, mutual adaptations can sometimes be left to automatic processes; but other times, as I indicated in essay 3, interventions are needed either to restore these processes or to intervene so that better, rather than worse (socially parasitic or predatory), adaptations develop. This leads to a third answer: Yes, some social adaptations are better than others. And, as I have argued in *Philosophy as Diplomacy* and shall explain in this book, the answer I just gave implies that adaptations can be assessed objectively, at least in that reasons can be given for or against them. Indeed, this is all that is needed for the nonarbitrary critical scrutiny of alternative adaptations. It is not incompatible with the existence of an independent, unchanging standard for assessing alternative adaptations; but, as I shall explain later in the book, it does not require it.

— I can't wait to see you explain all these things! Perhaps then I will be able to tell whether your position agrees or disagrees with the deep ecology movement.

— This is a question I have asked myself. However, I have difficulty answering it, partly because of the various current conceptions of deep ecology, partly because their applications are often concerned with long-term objectives and ideals rather than, as I am, with processes of social change here and now. To put it another way, I am primarily concerned not, as the deep ecology movement appears to be, with long-term objectives and ideals. I am concerned with how to deal with problems and issues in which the aims themselves are at issue and in which a main concern is how to go about settling the issues and the aims at issue in an effective and morally sound manner within the time available and under existing economic, political, cultural, and other constraints.

—I am beginning to get lost in generalities again.

— Maybe it will help you to consider my application of the ecological model to environmental issues and the related engineering and technology culture and business culture.

— Maybe. Please proceed.

PART 2

Ecodiplomacy:
Ethics, Policy Making,
and Environmental Issues

ENVIRONMENTAL ISSUES

In August 1984, at the Mexico City United Nations International Conference on Population, the United States delegation prompted controversy at home and abroad by announcing that it would withhold aid from organizations that promoted abortions, thus endangering the funds it provided to the International Planned Parenthood Federation through the U.S. Agency for International Development.[1] The U.S. demand for assurances that abortion programs would not be promoted was met to the U.S. delegation satisfaction, and a conference report was adopted by consensus, but not without further controversy and a two-and-a-half hour debate about the propriety of including an Arab-supported resolution, which implicitly condemned Israeli settlements on the West Bank of the Jordan River.[2]

In October 1984, controversy erupted in Canyonlands, Utah, over the then-likely prospect that at least one of two nearby sites might be selected, in accordance with the 1982 Nuclear Waste Policy Act, to store radioactive wastes expected to remain toxic for at least one hundred centuries.[3] In December 1984, the Energy Department picked sites in Texas, Nevada, and Washington as the leading candidates to be the first U.S. permanent burial ground for nuclear waste, but it immediately ran into legal and political opposition and was submerged into controversy about the matter, a conflict still raging years later.[4]

In December 1988, Chico Mendes, the president of the Xapuri Rural Workers Union, who had organized rubber tappers to negotiate with loggers who were slashing and burning the rain forest and thus making room for cattle ranchers, was murdered by members of the cattle rancher group. This murder led to an escalation of the—still ongoing—conflict, which involves participants in Brazil and abroad.[5]

In February 1989, a settlement between Union Carbide and the Indian government, whereby Union Carbide agreed to pay $470 million for the 1984 Bhopal methyl isocyanate leak that killed more than 3,300 people and injured tens of thousands, sparked controversy on the grounds that the victims had been betrayed. This led to a never-ending legal and political nightmare whose end, one decade later, was still not in sight. Indeed, the $470 million that, adding interest, by then amounted to almost $700 million, was still sitting in the State Bank of India.[6] Conflicts such as these are central constituents of *environmental issues*, which are characterized by sharp differences of opinions or conflicts of demands about environmental concerns and their apparent, or real, conflicts with economic, political, technological, or cultural concerns, together with what people do to uphold their opinions or satisfy their demands.[7]

The concerns involved in these issues have prompted a variety of responses in the form of national laws and international agreements. For example, in the United States, the responses typically took the form of polluter-pays provisions embodied in the Clean Air Act and its 1977 amendments, the Clean Water Act, the Solid Waste Disposal Act, the Toxic Substances Control Act, the Surface Mining Control and Reclamation Act, and other pieces of legislation. At the international level, a variety of agreements were reached, ranging from those signed at the UN Conference on the Environment held in Stockholm in 1972, through those signed at the UN Conference on the Human Environment held in Nairobi in 1982, to those signed at the 1992 UN Earth Summit held in Rio de Janeiro, and to the blueprint signed at the 1994 UN International Conference on Population and Development in Cairo.

Attempts of this sort have been increasingly emphasized in environmental policy making since the 1970s, prompting a great deal of support among some, and many criticisms among others. This raises a variety of questions: Do these primarily legalistic attempts work? How helpful are they in addressing environmental issues? Does soundly addressing these issues call for a shift in emphasis, away from an exclusive or primary focus on legislation and treaties? What alternative and plausible approaches are there? These are the main questions I will address in this essay.

I will advance six theses. First, national legislation and international treaties have attained and are likely to attain their purposes only to a limited extent and in a mixed manner. Second, they have not been and, unless current circumstances change both in the United States and abroad, are likely not to be soundly implemented in the foreseeable future. Third, in some—typically international—cases, their implementation shortcomings result from lax enforcement, lack of commitment—especially concerning who is going to foot the bill—and lack of human resources. Fourth, in other cases—for example, in U.S. cases—the shortcomings result from overzealous but misguided enforcement, often because of an exclusive focus on certainty or near certainty of risk avoidance and on adversarial approaches. Fifth, these shortcomings should be addressed not merely by trying to increase commitment at the international level and improve and redirect enforcement at the national level, but also, and primarily, by focusing on such social decision procedures as negotiation, bargaining, mediation, arbitration, consensus building, and various forms of convention settling. Finally, past uses of less legislation-dependent procedures, as well as current policy-making developments, evidence a greater emphasis on such approaches, hence provide evidence for their political feasibility.

Since environmental issues and the alternative environmental policy-making options about them affect or are likely to affect people's lives and society at large, the policy-making problems posed by the questions and addressed by the theses just formulated are also ethical and sociopolitical problems. Accordingly, in addition to relying on environmental policymaking literature, I will approach the problems against the backdrop of contributions that ethical and sociopolitical philosophy can make to the resolution of the problems. In order to apply these contributions realistically, I will begin by examining some cases of environmental legislation and implementation.

LAX ENVIRONMENTAL COMMITMENTS AND ENFORCEMENT

International Lassitude

Environmental concerns clashed with other concerns in September 1994, at the Cairo United Nations International Conference on Population and Development. Just as they had in 1984 and 1974, they clashed because of fear on the part of the Vatican and other countries (which, this time, did not include the United States) that the final document would imply, however indirectly, the permissibility of abortion.[8] At the conference's end, a nonbinding twenty-year blueprint was signed by one hundred eighty countries, even with partial Vatican support; but coming up with the implementation funds was left up to each individual country.[9]

Given this lack of financial commitment, the blueprint was a success more for its promise than its achievements. So had been the 1992 Earth Summit nonbinding Declaration on Environment and Development, and the Agenda 21 whose implementation cost was estimated to be $125 billion a year; the legally binding Global Warming Convention, which was signed by 153 countries but did not set targets for reducing carbon dioxide emissions; and the legally binding Biodiversity Convention, which was signed by 153 countries, but not by the United States, though the Clinton administration later announced that the United States would sign it.[10]

Leaving all, or nearly all, such financial matters to be worked out in the future or by parties that, in some cases, are very poor is a double-edged sword. As many an international agreement evidences, it facilitates agreement on principles and guidelines but opens the door for significant implementation failures. Let us take a look at the limited success of UN environmental action since 1972 and at why this success has been limited.

By 1992, UN reports indicated that the environment had worsened in the previous decades. The 1972 Stockholm Action Plan had been only partially implemented, and the results had been dismal. One might argue that the limited success of the Stockholm Action Plan was caused not so much by lack of financial commitment but by lack of support on the part of developing countries. These countries were suspicious that environmental proposals were aimed at undermining their economic development and that a global network of air monitoring stations would become bases for subversion.[11]

But the suspicion was gone in Nairobi in 1982. Instead, whether due to a change in world economic conditions, a change in industrialized countries' policies, or both, wealthier nations that had taken the lead in Stockholm reduced their commitment of resources toward resolving global environmental problems. For example, since 1972, the United States had been contributing about $10 million to the United Nations Environment Programme each year. But, in fiscal year (FY) 1982, the Reagan administration requested only $2 million. The amount was increased to $7.85 million by Congress, but the Reagan administration worked to reduce it again in FY 1983, arguing that it was time for others to pick up more of the tab and that natural market forces should be given more of a chance to help resolve the problems.[12]

In this regard, the final Nairobi declaration also included some provisions that, at the time, were seen as controversial. One of them called for a new international economic order.[13] This view was echoed, and no longer controversial, by Brazil's president at the opening of the 1992 Earth Summit. Hence, there had been at least attitudinal and intellectual progress. But at the end of the summit, financial commitments were more promise than reality, and matters concerning specific targets, timetables, monitoring mechanisms, incentives, and sanctions were largely left open for future negotiation.[14]

In response, it can be argued that this approach is only sensible, because it would be impossible to work out the myriad of details involved during the short span of a world summit. Yet, the previous review of UN environmental programs during past decades gives reason for cautious skepticism concerning the effectiveness that, in and of themselves, international agreements about environmental matters are likely to have. This is not to say that they are unnecessary. It is simply to say that they are merely opening moves in a policy-making process that, as I will further argue, can hardly succeed along exclusively legalistic lines.

The shortcomings just described might be thought to be merely enforcement failures. But this is an exaggeration. No doubt, enforcement failures exist. But implementation failures occur, and are likely to continue, even when enforcement is effective. This is best established by approaching environmental policy implementation at the level of national policies and national policy implementation.

National Lassitude

Case 1: Argentina's National Parks

One might think that the implementation shortcomings discussed in connection with international treaties occur when, in particular countries, the treaties are perceived to be in conflict with these countries' interests. Yet, there is evidence to the contrary. For some countries display environmental policy implementation failures even concerning matters that, judging by their national environmental policies and the degree of national consensus reached about them, are in the countries' interest.

Consider, for example, national park policy implementation in my own country, Argentina, where there has been substantial consensus supporting national parks for many decades. According to this policy, Argentine national parks are subject to regulations analogous to those of U.S. national parks. One difference, however, is that in the United States the protected areas are two-and-a-half times greater (relative to its territory)—and seven-and-a-half times greater in absolute values—than those in Argentina, a country whose area is about one-third that of the United States.

This evidences a different degree of preservationist or conservationist commitment, or at least awareness, between these countries. But the differences are not merely attitudinal or a matter of knowledge. A second difference reflects the (until recently) highly regulatory functions of Argentina's government. In the United States, there are substantial private attempts at environmental protection—including such things as land purchases and public education—that supplement those of the National Park Service. By contrast, in Argentina, these attempts are limited and quite recent.

There is also a difference that is neither a mere matter of attitudes or knowledge, nor the result of a vacuum produced by a regulatory past. It also concerns political practices (which, of course, involve attitudes, but are not exhausted by them): Environmental law enforcement—like all law enforcement—in Argentina is extremely lax. The result is that, out of twenty-one species of mammals, five species of birds, and one reptile species that are known to have become extinct in areas designated as national parks, many of them became extinct *after* the areas became national parks.[15]

If these factors can be, and, as evidenced, sometimes are, present concerning policies that reflect a country's national interest, there is reason to believe that they can and are likely to be present concerning the implementation of international treaties, even when no national interest is perceived to be in conflict with the treaties.

One might be tempted to reply that better enforcement coupled with financial assistance should suffice to deal with the implementation failures just described. But this is too simplistic. For first, as indicated, the said law enforcement failures are dependent on practices concerning the legal systems within the scope of which enforcement is supposed to occur. This *customary law* element cannot be eradicated overnight, or even within the time available for dealing with environmental issues.

Second, human resources are also a crucial variable. Eduardo H. Rapoport estimated in 1990 that there were ten thousand endangered species in Argentina. Even if only some of these were to be protected, some order of priorities would have to be established to decide which ones to protect. For this, each species's situation would have to be studied. That is, at least ten thousand research projects would be needed, unless the study of the factors endangering some of these species would also involve the study of the factors endangering some of the other species. In any case, the studies needed are likely to be quite a few. But, in 1990, Argentina had about eight hundred ecologists (biologists, agronomists, veterinarians, and other professionals), only a few of whom devoted their efforts to conservation and genetic resources. In other words, even with money available, sensible park size selection, and goodwill on the part of all affected, the human resources were not available to carry out the studies likely to be needed in the environmental protection task faced. Argentina's environmental predicament was not encouraging.[16]

Judging by the fact that, even counting all its economic, bureaucratic, and other woes, Argentina is not close to being one of the poorest countries on earth, its environmental predicament is likely to be shared, perhaps in more dramatic fashion, by other countries. Yet, even in countries where law enforcement is often successful, implementation failures do occur. Let us, in this regard, consider a U.S. case.

Case 2: New Mexico's Bitter Lake Wildlife Refuge

In November 1982, Bob Burnett, New Mexico Wildlife Federation's vice president, was in the vicinity of the Bitter Lake National Wildlife Refuge when he noticed heavy equipment inside the refuge area. It turned out to be drilling equipment owned by Yates Petroleum, a company that, without any state or federal permits, had been drilling in the refuge area with full governmental knowledge of its trespass. Informed of the situation, the Department of Interior took ten days to act and then seemingly only because public outcry was so loud.[17] This enforcement failure, coupled with the Department of Interior's reluctance to respond quickly to the matter, evidences that the event was not an aberration but that there was a tendency toward enforcement lassitude when it came to enforcing environmental legislation on energy companies or, for some reason, at least on Yates Petroleum.

One might be inclined to argue that the said lassitude merely reflected the new policies of the Reagan administration. However, though this may be arguable when one focuses on enforcement failures, it is less convincing when one recognizes that enforcement failures are only one way in which environmental policy implementation can fail. The case of the Tellico Dam and the snail darter fish evidences this fact. Let us consider the circumstances surrounding it.

Case 3: The Tellico Dam

First, there is little question that the Carter administration—especially President Carter himself—took environmental concerns and related legislation seriously. Second, the snail darter fish was, at the time, thought to be an endangered species. For it was thought to inhabit only one cave, which happened to be in the way of the bulldozers in charge of finishing the Tellico Dam in Tennessee. The Department of Interior had little choice except placing the fish on the endangered species list. It could not redesign the fish's habitat in any way: Given the knowledge available at the time, it was just that cave. Only Congress could decide to make an exception to the application of the Endangered Species Act. However, the Tellico Dam project had always been controversial, even in Tennessee. Confirming this history, in February 1979, a cabinet-level committee created by Congress unanimously voted against the project.

Yet, President Carter needed all the votes he could get to have the Panama Canal treaty ratified, and these included the votes of Tennessee Senator Howard Baker and Tennessee Congressman John J. Duncan. These legislators had succeeded in tying the completion of the dam to an appropriation bill for energy and water projects and, faced with the prospect of losing Panama Canal treaty ratification, President Carter signed the bill into law on September 25, 1979.[18] The point is not whether the Endangered Species Act is justified. I do not intend to pursue this line of argument here. The point is, rather, that policy implementation failures can occur in ways other than enforcement failures, for example, as in the Tellico Dam case, through pork barrel legislation.

The previous discussion might lead one to believe that environmental policy shortcomings are merely a matter of lack of resources, weak enforcement, and morally indefensible politics. But they are not. Some of the failures are failures of technological assumptions and political conventions that are largely a matter of policy-making culture, when not fads. I will examine these next.

EXCESSES OF MISGUIDED ENFORCEMENT

Some technological assumptions and political conventions involved in environmental policy making lead to the mismanagement of time, energy, and financial and human resources that could be better used to pursue environmental goals. This was dramatically exemplified in 1990. After a ten-year effort to force cleanup of a toxic waste dump in southern New Hampshire, the forty-thousand-page record, with which all parties involved seemed to agree, stated that the waste dump, which was a swamp—hence certain not to have children playing on it—was clean enough for those nonexistent children on the site to eat small amounts daily for seventy days each year without significant harm. Also, through natural evaporation, it was predicted to have, at most, half of the contaminants currently in the dump by the year 2000. Nonetheless, enforcement-prompted litigation about the case made it to Chief Judge Stephen Breyer's U.S. Court of Appeals for the First Circuit. This was because, as Breyer reports, the only private party that had not settled sued not to spend the $9.3 million needed to burn the soil—as still demanded by regulators—so that the nonexistent dirt-eating children could eat small daily amounts for 245 days per year without significant harm.[19]

One might think this latter example is a legal, administrative, and financial oddity. Yet, available evidence suggests otherwise. According to a 1992 study, about 90 percent of insurers' and 20 percent of liable firms' expenditure on Superfund environmental claims were not on actual cleanup but on "transaction costs," most of which were legal fees.[20] A 1991 study concludes that a conservative estimate of transaction costs is 10 to 15 percent of direct cleanup costs, or about $4 million average per National Priority List site.[21] Such financial waste, mostly in the form of legal fees, was significantly the result of administrative and legislative constraints, tied in to tunnel vision approaches to risk or harm, which deserve close scrutiny. I will turn to these next.

HARM, RISK, AND POLICY-MAKING TUNNEL VISION

An old Yiddish proverb says, "Ever since dying became fashionable, life hasn't been safe." This basic truth prompts an enormous variety of approaches to risk through environmental policy and decisions. These are not merely imaginary approaches concocted by out-of-touch philosophers in their armchairs and quiet halls of academia. On the contrary. They have become so customary and generalized throughout entire societies or societal sectors that they sometimes are at the core of what amount to policy-making subcultures. In what follows, I will examine a few predominant types of policy-making subcultures in an attempt to clarify their roles and policy-making status in environmental issues.

Pareto Optimality

A significant policy-making social sector in the Unites States and abroad adopts what can be called the *Pareto optimality approach*. It is evident in the position of many influential policymakers—frequently economists—who favor only Pareto efficient policies and decisions, that is, only those which, according to good available evidence, will help some while harming no others.[22] However, it is not self-evident—indeed, it is arguably false—that such an approach is justified. As Aaron Wildavsky points out, Pareto optimality "can strangle growth and change, because any new developments are likely to hurt someone, somewhere, sometime."[23]

One might question Wildavsky's view that it is impossible to separate harmful from beneficial effects. Still, there are policies and decisions that would do away with such substantial disadvantages, which many suffer, that, arguably, are worth seeking despite the loss to others. For example, protecting the civil liberties of oppressed individuals or groups is justified even if doing so is greatly irritating and obstructive to those who oppress them or take advantage of these individuals' oppression.

Analogous evidence can be found concerning environmental issues. For example, policies and decisions aimed at protecting the Brazilian rain forest rubber tappers against cattle ranchers are arguably justified even though greatly irritating and obstructive to the cattle ranchers. Such policies and decisions would protect the rain forest, hence the interest of humans, whose survival would be endangered by weather-pattern changes resulting from rain forest destruction.

The preceding discussion provides reasons for concluding that Pareto optimality is not only not sufficient, but also not necessary for policy and decision to be justified. As Charles E. Lindblom puts it, "There remains a conflict as to what is to be done."[24] Let us now turn to other approaches to risk in environmental policy making.

Guilty Until Proven Innocent, or Whenever in Doubt, Abstain

Many individuals and groups involved in environmental issues who seek assurance against risks will show total support for policies and decisions that, for example, protect rubber tappers against cattle ranchers in the Brazilian rain forest. Their reasons for this position, however, amount to sheer risk aversion, which, like Pareto optimality, can undermine change for the better and, indeed, changes toward greater safety for humans and nonhumans. At the core of this position is the view that environmental policies and decisions should rule out any activity—from the development of a new sea resort to the use of a new tiller in agriculture—until there is evidence *conclusively* establishing that no adverse environmental consequences will result from these activities.

Those who hold this view are set on avoiding not just all likely risks but all possible risks that have not been shown to be utterly unlikely. In other words, all activities associated with possible risks—and, in life, that means all activities—are treated as *guilty until proven innocent, or in accordance with the rule: whenever in doubt, abstain.*

One might question the operant character of this approach. But it is evidenced, for example, in bills introduced to regulate the release of genetically engineered organisms by requiring that those intending to do so "must be able to prove that no adverse effect to the environment will occur as a result of these actions."[25] In addition, the approach is applied not only to risks associated with activities not yet actualized—say, those involved in implementing a new technology—but also to risks associated with past activities—say, as in the New Hampshire swamp previously discussed, those associated with wastes resulting from past practices and technology uses. Along these lines, under United States Superfund legislation, the Environmental Protection Agency (EPA) is supposed to clean up toxic waste sites and, if all other measures fail, to truck the wastes to secure disposal sites where pollutants cannot escape into the environment. But since no such secure site has been invented or discovered, and EPA inaction is not permitted by the legislation, millions of dollars are being spent to transport wastes from one insecure site to another.[26]

This criticism is in no way meant to assign a monetary value to health risks associated with waste—a topic on which I intend to make no pronouncements in this essay. It simply points to the fact that, given current and foreseeable circumstances about waste-disposal technology, and given that money, time, and human resources are limited, these are being badly mismanaged and ecologically minded aims are being undermined as a result of such tunnel vision, often highly risk-avoidance prone legislation.

In response to criticisms such as this, weaker views have been advanced concerning what and how unlikely the risks are supposed to be. I will examine a common one next.

No Risk Not Proven Quite Unlikely, Small, Promptly Recognizable, and Correctable

In discussing nuclear power plants, Robert E. Goodin argues that, in the case of nuclear reactors, we must have good reasons for believing that whatever errors a policy allows to occur in their construction and operation, if they occur, will be small. For large errors may lead to catastrophe. Even small errors must be promptly recognizable and correctable. For "the impact of radioactive emissions from operating plants or of leaks of radioactive waste products from storage sites upon human populations or the natural environment may well be a 'sleeper' effect that does not appear in time for us to revise our original policy accordingly."[27]

I sympathize with Goodin's concern. The world is, indeed, an unsafe place, and it is good to prevent risks, especially serious risks. But we should be careful with the harm and new—sometimes worse—risks likely to come from some ways of preventing risks. In this regard, Goodin's criteria are too strong and likely to backfire. This was noticed by Aaron Wildavsky, who, in criticizing Goodin's views, made two points. First, despite current understanding of radiation, no one can reasonably claim that any particular dose of radiation, however small, would not irreversibly damage any particular individual or population at any given time. Second, the long-term effects of small radiation doses are not well known yet.[28] But this lack of knowledge, though perhaps a very good contributing reason, is not, by itself, a decisive reason to outlaw the construction and further operation of all nuclear power plants without any further ado. One must also consider the risks involved in alternative energy-related technologies—say, in those powered by fossil fuels—the social changes involved in their implementation—say, their effects on labor and other affected populations—the constraints—such as the knowledge limitations—under which policies and decisions on the matter must currently be made, and the risks involved in trying to overcome these knowledge limitations.

As for how to overcome knowledge limitations, Goodin himself acknowledges that "learning by doing . . . has been . . . responsible for dramatic improvements in the operating efficiency of nuclear reactors."[29] However, this is no reason for recklessness. The question is whether the greater safety that resulted from running such risks justifies having run them. One might be tempted to say it does not justify it because, in running such risks, we would still be living with risk, and all risk should be avoided. Yet, this reply is too strong and unnecessary. In fact, it simply reverts to the previously discussed "no risk until proven totally unlikely" position on which, presumably, Goodin wishes to improve. His answer is more cautious: "We would be living not merely with risk but also with *irresolvable* uncertainty."[30] Despite its cautiousness, however, Goodin's answer has some significant shortcomings. Let us examine them.

First, Goodin's answer conflicts with his previously stated criteria. For his requirement that we "have good reasons for believing that the errors, if they occur, will be small" and "immediately recognizable and correctable" makes room for uncertainty. After all, good reasons need not amount to conclusive proof—the type of thing that would eradicate uncertainty. But Goodin's requirement that we avoid living with irresolvable uncertainty, requires that our reasons *make certain* that the risks run are small, promptly recognizable, and correctable.

Second, this approach amounts to making it totally unlikely that the risks run, if any, are anything other than small, promptly recognizable, and correctable. To be sure, this is not the "no risk until proven totally unlikely" view. It is the "no risk until proven totally unlikely to be anything but small, promptly recognizable, and correctable" view. Unfortunately, this view is open to criticisms analogous to those advanced against the more general version. For, as Wildavsky points out, " every technology, viewed in advance, has 'irresolvable' uncertainties."[31] Not just space and aeronautic technology, but a variety of technologies to which we are accustomed—from wood burning and agriculture, through water navigation and the automobile, to natural gas heating—initially had, and still have, some unresolved uncertainties that, given the constraints of time and the finitude of human lives, are irresolvable uncertainties with which many must live.

Indeed, Goodin's position implies that not even solar power technology should be implemented, because the available evidence does not make it totally unlikely that the associated risks of heat pollution and inadequate energy supply will be anything but small, promptly recognizable, and correctable. That is, if solar power turns out to be, as current evidence—not without, for some time to come, irresolvable uncertainty—suggests, the overall safest technology to humans and nonhumans, Goodin's approach would preclude policymakers from taking appropriate action to implement it now. The opportunity would be at least postponed, and probably lost, because of our inability to ensure that no irresolvable uncertainty will accompany such implementation. This way, by taking fewer risks, one would decrease, rather than increase safety.

In addition, since current evidence shows no other—say, eolic—technology to be clearly freer from risks than solar, then Goodin's approach would preclude the implementation of any other technology as well. This is absurd and shows the policymaking untenability of the approach and its implied centerpiece, the "guilty until proven innocent" or "whenever in doubt, abstain" rule. In other words, any environmental policy and decision based on such rule is bound to be flawed.

One might respond that the previous criticisms rely on too unsympathetic an interpretation of Goodin's position by putting too much weight on Goodin's use of the phrase "irresolvable uncertainty." One might further argue that Goodin did not really mean to take such a strong position, but only to say that we should always have good reasons for believing that the risks we run are small, immediately detectable, and correctable.

In response, I must say that, in discussions such as the present one, using the phrase "irresolvable uncertainties" without meaning it is not good practice. Besides, the suggested weakened interpretation is still too strong. For, even in the case of solar energy, it is debatable that the evidence's suggestion that solar energy will be free from heat pollution and other side effects amounts to good reasons. The main problem with Goodin's approach is that it seeks all answers beforehand, when, in fact, some of the answers and the reasons can only be worked out through an ongoing process, and policy-making decisions must take place in the meantime.

No doubt, Goodin's position is sound in entailing that we should keep our options open. This is what we should do in engaging in the process just mentioned. But our keeping our options open should not depend on conclusive probabilistic evidence concerning the particular technology or technologies at issue at decision time. Instead, it should depend on the fact that, as we well know from past policy-making experience, initially unpredictable, yet significant developments are never entirely out of the question. For example, deaths and injuries resulting from motor vehicle accidents in the United States in the late twentieth century resulted partly from events such as the construction of the U.S. interstate highway system in response to the cold war, which was totally unpredictable at the time motor vehicles were introduced in this country.

A BEGINNING TOWARD ENVIRONMENTAL POLICY-MAKING BALANCE

Curbing Technological Immodesty and Policy-Making Tunnel Vision

The matters just discussed are not the only ones affecting environmental legislation. Sometimes they are, purely and simply, technological limitations. But a certain kind of technological unawareness or immodesty has made it very difficult to acknowledge them. After spending billions of dollars on some of the United States' worse toxic sites, the EPA threw up its hands claiming that, given 1992 state-of-the-art technology, it would take thousands of years to clean groundwater contaminated with solvents.[32]

In addition to technological shortcomings, ignorance of environmental legislation's effectiveness is often part of the picture. For example, by the spring of 1992, when the reauthorization of the 1987 Clean Water Act was already overdue, U.S. databases presented a fragmented picture of national water quality. Almost no information about potentially toxic organic compounds and trace metals was available.[33]

These examples provide additional evidence for concluding that however wise environmental legislation may become, its implementation has been and, in significant respects, is likely to keep on being limited. Indeed, since its limitations largely result from tunnel vision, as well as technological and scientific shortcomings, its limited effectiveness is not exclusively dependent on polluter-pays provisions in such legislation or on public resistance to such provisions.

The various reasons formulated so far support this essay's first four theses: First, national legislation and international treaties have attained and are likely to attain their purposes only to a limited extent and in a mixed manner. Second, they have not been and, unless current circumstances change both in the United States and abroad, are likely not to be soundly implemented in the foreseeable future. Third, in some—typically international—cases, their implementation shortcomings result from lax enforcement, lack of commitment—especially concerning who is going to foot the bill—and lack of human resources. Fourth, in other cases—for example, in U.S. cases—the shortcomings result from overzealous but misguided enforcement, often because of an exclusive focus on certainty or near certainty of risk avoidance and on adversarial approaches.

To support these theses is not to say that environmental laws and treaties are unnecessary or that environmental safety is unimportant. On the contrary. Environmental laws and treaties are only a first and partial move in dealing with environmental issues. And safety is crucial; but it should be addressed in a manner that is balanced, based neither on desperate demands for safety with certainty, nor on careless disregard for risks however likely and catastrophic the risks involved. I will next outline some features of this balanced approach.

The Place of Safety Approaches in Policy-Making

Rejecting the policy-making approach based on such rules as "guilty until proven innocent," or "whenever in doubt, abstain" may land us in the opposite, and equally foolhardy approach based on the rules of innocent until proven guilty or whenever in doubt, take any risks. To be sure, this latter policy-making approach would make room for risky activities only so long as no good evidence becomes available indicating that they are serious, perhaps irreversible. However, without any further constraints, this position would open the door for a great deal of reckless implementation of activities and technologies that have hardly been studied or tested.

In this regard, Wildavsky's criticisms of Goodin's and other anticipatory approaches to risk and, by contrast with these, his defense of resilience-building approaches, such as trial and error, and the net-benefits criterion can be misleading, if not flawed. To be sure, his purpose "is not to deny the importance of taking steps to lessen technological dangers but, rather, to put these dangers into perspective."[34] But there are small risks and there are large risks; and it matters very much whether risks are run voluntarily or involuntarily, how they are distributed among the population, whether they affect ordinary people or only specially sensitive people (as in the case of those allergic to fragrances), whether the risks are clear and present or unclear and future (and, if so, how delayed and how spread), whether the sources of the risks are likely to be misused, and whether those running greater risks are somehow compensated for it.[35] Wildavsky considers some, but not all, items in this certainly nonexhaustive list of relevant factors. This slants his analysis and undermines his laudable purpose of putting the dangers in perspective.

One obvious—but incomplete—corrective is to require, as previously indicated, that the activities and technology being considered be reasonably studied and tested before proceeding to implement them, first in a controlled and limited manner, until reasonably good enough evidence—which may, but need not be, conclusive—is gathered, indicating that the activities or technology's generalized use is likely not to involve serious dangers that are not outweighed by any other consideration. But whether they are outweighed or not is not a matter that can be settled in advance. In order to consider the other factors listed above, some discussion and, sometimes, limited and controlled interaction of the proposed activities and technology with those likely to be affected is crucial. In addition, such social decision procedures as bargaining, negotiation, mediation, arbitration, consensus building, and various forms of convention settling are also crucial.

A crucial consideration here is not whether outcomes will overall be desirable or fair—which may not be determined in advance—but whether the constraints on the procedures just mentioned make them both effective and morally acceptable. For example, it will be crucial whether those affected are reasonably represented in the process and whether there is room for corrections sensitive to these people's interest.

Making room for such process points to the more cautious position that advises *move slowly, incrementally*. There are many varieties of this approach. A crucial point common to them is, as Lewis Dexter wrote in a letter to Wildavsky, that even if one does not take invariably fewer or, perhaps, smaller risks—which, as previously indicated, can reduce safety—one can still improve one's prospects by "going slow frequently."[36]

Since overall systematic accounts and sufficient evidence to assess these risks with absolute precision are unavailable and at issue, incremental policy-making approaches are, at least in general, preferable. However, there are those who still object to the generalized use of incrementalism because grand opportunities, or policies and decisions that require long-term support—say, the space program, and the genome project—would find no place in it. As a result, policy making would not make room for a variety of worthwhile technologies. For example, it would not lead to space technologies used to predict weather patterns or—as was the case when I was in north-central Florida in 1982–83—to space technologies that, by providing data on the unusual and expanding browning of treetops, helped discover unknown underground concentrations of pollutants that had reached the north-central Florida aquifer.[37]

This objection does not in any way show incrementalism to be inapplicable or useless in environmental issues. Yet it does show that either it is not always sound or, at the very least, its scope of applicability is controversial. That is, it is an approach that, however wiser than the others, is not above controversy, but *at issue* in environmental issues.

One might object that such a fact is irrelevant for assessing approaches to safety in and of themselves. This may be true, but it is irrelevant to environmental policy making, where the fact is that, as I previously argued, a variety of approaches to safety, however unwise, are operant. Further, as previously argued along the lines of Lindblom's criticism of Pareto optimality, the fact that various approaches are at issue evidences that there is an actual conflict about what ought to be done, and this is part of the issue to be addressed by environmental policy making.

One might respond that, though perhaps politically relevant, this is a merely political circumstance, which has nothing to do with the concern, expressed at the outset, with environmental philosophy or the moral evaluation of environmental policies and decisions. But these circumstances are all too real. Disregarding them would be blatantly unrealistic, hence irrational. I will next address this objection.

ETHICS, SOCIOPOLITICAL PHILOSOPHY, AND ENVIRONMENTAL POLICY MAKING

Consider the case of Chico Mendes, the rubber tappers, and the loggers in Brazil again. Policies and decisions aimed at protecting the Brazilian rain forest rubber tappers against cattle ranchers—hence the rain forest, hence the interest of humans whose survival would be endangered by weather-pattern changes resulting from rain forest destruction—are arguably justified even though greatly irritating and obstructive to the cattle ranchers. However, this does not establish that unilaterally protecting the rubber tappers is a sound way of addressing the issue, but only that there is a conflict about what ought to be done and that, were one to abstract from everything else, some moral and prudential considerations would warrant unilaterally protecting the rubber tappers.

Yet, as indicated at this essay's outset, in dealing with the issue, the conflict between rubber tappers and loggers and, indeed, cattle ranchers must be addressed. This is precisely what Chico Mendes saw. Granted, principles of justice and even worldwide prudential considerations seem to have been on his side. But he knew better than simply relying on these alone to address the issue. Instead, he used a variety of social decision procedures, most notably, the *empate*. Lisa H. Newton and Catherine K. Dillingham describe it as follows:

> The union's primary tactical weapon was the *empate*, a species of nonviolent confrontation developed at the beginning of the battle to save the Amazonian Rainforest. An *empate* is conducted like this. Once the location of future logging was determined, the union leaders would enlist tappers who then marched to the site. There, after a civil conversation with the loggers, which might include an assessment of the loggers' future after the trees were cut, the loggers usually gave up their chain saws peacefully and went home while the tappers destroyed their camp. (The loggers and tappers had a lot in common as victims of bosses.) Each empate set back the ranchers' progress and absolutely infuriated them. The rubber tappers actually managed to drive two of Brazil's largest ranchers out of the area.[38]

The *empate* is one of many social decision procedures whose effectiveness and moral justifiability must be assessed in establishing how to deal with the issue. It is a form of side stepping that worked in the Amazonian rain forest case so long as the situation did not become predominantly confrontational. Here, ethics and sociopolitical philosophy can make a contribution. It can, for example, help deal with the ever-present questions: Is this social decisions procedure—say, the *empate*—justified in the situation being faced? Is the situation primarily a controversy or primarily a confrontation? What other social decision procedures are likely to be not merely effective but morally justified in the circumstances? These and related questions can be addressed through the conception of philosophy as diplomacy, which I have developed elsewhere and outlined in the previous essays.[39]

At this point, another, more practical objection could be raised. One might argue that, even if the philosophy as diplomacy approach works at some level, it is too general. What is needed are viable models for environmental policy making, and the preceding discussion, however enlightening, has offered none.

In response, two points need to be made. First, the philosophy as diplomacy approach needs to be general because it covers a great variety of social problems and issues. Second, as the objection suggests, there is a need for viable models applicable to environmental policy making. These, however, need not, and, in fact, do not, conflict with the overall philosophy as diplomacy approach. I will propose one such model in the next section.

ENVIRONMENTAL MANAGEMENT PARTNERSHIPS

The limited effectiveness often attained by environmental legislation and treaties in the face of regulatory lassitude and other limitations is frequently of an extralegal and extra-administrative nature. To be sure, at least in the United States, the legislation gave negotiation and bargaining power to environmentalists. For example, the National Wildlife Federation brought a lawsuit against the Department of Interior in the early 1980s on the grounds that the department's planned weakening of forty-six strip-mining controls would potentially harm the environment. As a result, the Department of Interior agreed to delay the changes until environmental impact reports were filed. Whatever its final outcome, this case evidences how environmentalists learned to use the challenge process to win commitments from a variety of institutions, often in the form of negotiated settlements to the dispute. In addition, these settlements were often very specific about what the institutions would do to address the environmental concerns formulated.[40] And sometimes, as in the 1982 Michigan waste cleanup settlement, they were hailed by the EPA as a model for future settlements in the United States.[41]

Representatives of business or other interests have often charged that these settlements were forms of blackmail that undermined free enterprise. This response, however, is too abstract to address the actual environmental policy and decision problems in a politically effective and morally sound manner. For it disregards the actual conflicts mentioned at this essay's outset that prompted the enactment of environmental legislation, the efforts aimed at enforcing this legislation, and the negotiated attempts at dealing with the conflicts outside of the enforcement sphere.

A more defensible response is to regard the negotiated settlements as quick and fair ways of trying to resolve pending disputes. It is not clear, however, that they are invariably effective. For the settlements are parasitic on challenges, hence on adversarial approaches to dealing with environmental noncompliance and, during periods of enforcement lassitude or in countries where environmental legislation is weaker (or absent) and its implementation limited, there is less, or no, legal leverage for negotiating similar settlements.[42]

The limitation just mentioned is a reason for seeking still other—less legislation-dependent—approaches. Hence, it is also a reason to move away from an exclusive emphasis on legislation or international treaties, which, as argued so far, are needed but, by themselves, evidence a variety of policy-making shortcomings as tools for sound environmental policy making. This leads to our fifth thesis: These shortcomings should be addressed not merely by trying to increase commitment at the international level and improve and redirect enforcement at the national level, but also, and primarily, by focusing on such social decision procedures as negotiation, bargaining, mediation, consensus building, and various forms of convention settling.

One—though by no means the only—way of doing this could be to form *environmental management partnerships* between affected institutions and local communities concerned with their environment as well as with dwelling and making a living—say, by engaging in agriculture, strip-mining, or nearby fishing, or developing housing—in this environment. There is a precedent for this type of approach in the finance of community development that is, in some regards, of the urban environment. As it will be discussed in the next essay, some financial institutions and community organizations have had the wisdom to pursue this line of thought and action. Among the institutions that have been named as models for community development activities along these lines are South Shore Bank in Chicago, Illinois, and its Arkansas affiliate, Southern Development Bancorporation; the Center for Community Self Help in Durham, North Carolina; and Community Capital Bank in Brooklyn, New York. The Connecticut Housing Investment Fund, with People's Bank, whose headquarters are in Bridgeport, Connecticut, has developed and currently administers a home improvement loan program that targets low- and moderate-income homeowners.[43]

This suggestion has a precedent in private attempts at purchasing natural areas for conservation and preservation purposes. The difference is that there need not be any purchase involved, and the commitment of resources would be worked out between interested institutions and communities. Of course, having national legislation, international agreements, coupled with funds or economic incentives—say, tax breaks or lower interest rate programs—supporting the environmental management partnerships' activities would not hurt. Models for developing them are also available in community development programs that address urban blight and other urban environment problems.[44]

This short list is in no way exhaustive, but it offers a somewhat detailed idea of the models being used along the lines suggested in this essay. It also provides grounds for inferring our sixth thesis: Past uses of less legislation-dependent procedures, as well as current policy-making developments, evidence a greater emphasis on such approaches, hence, provide evidence for their political feasibility.[45]

The arrangements thus worked out are likely to be more effective and morally sounder than traditional, strictly adversarial approaches built in the polluter-pays acts the U.S. Congress enacted since the 1970s. For they will tend to prevent environmental deterioration, rather than simply to correct it—and, judging from past experience, to correct it very slowly and minimally—once it begins to happen. Such approaches could help us all deal more intelligently, and less parochially, with environmental issues. They also offer us humans hope of surviving ourselves and the consequences of our misdirected policies in these environmentally crucial times.

NOTES

1. Associated Press, "UN to Meet on Population: U.S. Criticized," *Hartford Courant*, Sunday, August 5, 1984, p. A5, col. 1.

2. "UN Conference Says Population Remains Challenge, Proposes Policies to Control It," *Hartford Courant*, Wednesday, August 15, 1984, p. D7, cols. 1, 2, 3.

3. T. R. Reid, "Nuclear Wastes Prompt Uproar in Canyonlands," *Hartford Courant*, Sunday, October 14, 1984, p. A13, cols. 1, 2.

4. From the *Washington Post*, "Protests Greet Choices of Nuclear Waste Burial Sites," *Hartford Courant*, December 20, 1984, p. A11. Combined Wire Services, "Toxic Dump Site Choices Challenged," ibid., Friday, May 20, 1986, p. A10, cols. 4, 5, 6.

5. For details of this case, see Andrew Revkin, *The Burning Season* (Boston: Houghton Mifflin, 1990), and Alex Shoumatoff, *The World Is Burning* (Boston: Little, Brown, 1990). For a philosophical discussion of this case, see Lisa H. Newton and Catherine K. Dillingham, *Watersheds* (Belmont, Calif.: Wadsworth, 1993), pp. 135–53.

6. Miranda S. Spivack, "Settlement Caps Bhopal Court Fight," *Hartford Courant*, Wednesday, February 15, 1989, pp. A1, A6; W. Joseph Campbell, "For Victims, the Ordeal Still Goes On," ibid., pp. A1, A6; Associated Press, "Bhopal Disaster Survivors Say Union Carbide Settlement Too Small," ibid., Thursday, February 16, 1989, p. A28; and W. Joseph Campbell, "Bhopal Blame Still Unclear," ibid., pp. A1, A29. For the fate of the $470 million, see Wil Lepkowski, "Union Carbide-Bhopal Saga Continues as Criminal Proceedings Begin in India," *Chemical and Engineering News*, Tuesday, March 16, 1992, pp. 7–14; and Sanjoy Hazarika, "Settlement Slow in India Gas Disaster Claims," *New York Times*, Thursday, March 25, 1993, p. A6.

7. For a more detailed characterization of issues, see my "Issues and Issue-Overload: A Challenge to Moral Philosophy," in *Philosophy as Diplomacy: Essays in Ethics and Policy Making* (Amherst, N.Y.: Humanity Books, 1994), pp. 1–12.

8. Associated Press, "UN Parley OKs Abortion Compromise," *New Haven Register*, Saturday, September 10, 1994, p. C5.

9. Associated Press, "180 Nations OK Population Control Plan," *New Haven Register*, Wednesday, September 14, 1994, p. A10. For the development of conference negotiations, see also Associated Press, "Population Conference Near Deal, Envoy Says," ibid., Monday, September 5, 1994, pp. A1, A4; "Vatican Still Stands Alone Against Abortion Document," ibid., Wednesday, September 7, 1994, p. A8; Associated Press, "Compromise on Abortion Unravels at UN Conference," ibid., Thursday, September 8, 1994, p. A6; and "UN Stays Out of Abortion Fray in New Plan," ibid., Tuesday, September 13, 1994, p. A5.

10. Associated Press, "Conference Highlights," *Hartford Courant*, Monday, June 15, 1992, p. A5. For the development of conference negotiations, see Stanley Meisler, "Economics, Politics Cloud Earth Summit's Goals," *Hartford Courant*, Thursday, May 28, 1992, pp. A1, A6; Associated Press, "European Environmental Chief to Boycott Summit," ibid., p. A6; "A World of Difference Since 1972 Conference," ibid., p. A6; Steve Nash, "On Eve of Earth Summit, Divisions, Doubts Persist," ibid., Tuesday, June 2, 1992, pp. A1, A5; Sam Dillon, "Earth Summit Polluted by Ill Will Toward U.S.," ibid., Thursday, June 4, 1992, pp. A1, A4; Eugene Robinson and Michael Weisskopf, "Germans Take Lead in Global Warning," ibid., Tuesday, June 9, 1992, pp. A1, A5; Sam Dillon, "Europe Snubs U.S. Views at Earth Summit," ibid., Wednesday, June 10, 1992, pp. A1, A5; "Earth Summit Sidesteps Issue of Birth Control, Activists Say," ibid., Wednesday, June 10, 1992, p. A5.

11. See, for example, I. Tangley, "UN Holds Global Environment Meeting," *Science News* 121 (May 29, 1982): 358. See also "A World of Difference Since 1972 Conference," *Hartford Courant*, Thursday, May 28, 1992, p. A6.

12. Tangley, loc. cit.

13. Ibid.

14. Combined Wire Services, "Earth Summit Ends on Note of Optimism," *Hartford Courant*, Monday, June 15, 1992, pp. A1–5.

15. Eduardo H. Rapoport, "Vida en Extinción," *Ciencia Hoy* 2, no. 10 (November–December 1990): 33.

16. Ibid.

17. J. Raloff, "Wilderness Trespasser Drills for Gas," *Sciences News* (November 20, 1982), p. 325.

18. Ross Sandler, "Tellico Dam," *Amicus* (fall 1979), pp. 4–5.

19. Stephen Breyer presents this case as an example of what he calls "tunnel vision" in *Breaking the Vicious Circle* (Cambridge, Mass., and London: Harvard University Press, 1993), pp. 11–12.

20. Jan Paul Acton and Lloyd S. Dixon, *Superfund and Transaction Costs: The Experience of Insurers and Very Large Industrial Firms* (Santa Monica, Calif.: Rand Corporation, Institute for Civil Justice, 1992).

21. John C. Butler III, "Superfund Super Costs," *Rethinking Superfund: It Costs Too Much; It's Unfair; It Must Be Fixed* (Washington, D.C.: National Legal Center for the Public Interest, 1991), pp. 67–77.

22. For two interpretations and a critical analysis of Pareto optimality in policy making, see Amartya Sen, "The Moral Standing of the Market," *Social Philosophy and Policy* 22 (spring 1985): 1–19.

23. Aaron Wildavsky, *Searching for Safety* (New Brunswick, N.J., and Oxford: Transactions, 1989), p. 18.

24. Charles E. Lindblom uses this example to defend not only the position I am taking —that, in some cases, eliminating great disadvantages to many may be worth the loss to some—but also the stronger view that securing the advantages to many may be so great as to justify the loss to some. See Charles E. Lindblom, "Who Needs What Social Research for Policy Making?" (paper presented at the New York Education Policy Seminar, Albany, N.Y., May 1984), cited in Aaron Wildavsky, op. cit., p. 231.

25. *Inside the Administration*, January 2, 1986. Cited in Wildavsky, op. cit., p. 14.

26. Robert E. Taylor, "In the Dumps: Toxic-Waste Cleanup Is Expensive and Slow and Tough to Achieve," *Wall Street Journal*, May 16, 1985, pp. 1, 20. and Cass Peterson. "EPA Urged to Shift Superfund Strategy," *Washington Post*, March 10, 1985, p. A7. See Wildavsky, op. cit., pp. 201–202.

27. Robert E. Goodin, "No Moral Nukes," *Ethics* 90 (April 1980): 418–19.

28. Wildavsky, op. cit., p. 21.

29. Goodin, loc. cit.

30. Goodin, op. cit., p. 418.

31. Wildavsky, loc. cit.

32. Mary Beth Regan, "EPA Has Given Up on Some Toxic-Waste Sites," *Hartford Courant*, Sunday, August 16, 1992, p. A2.

33. Debra S. Knopman and Richard A. Smith, "Twenty Years of the Clean Water Act," *Environment* 35, no. 1 (January/February 1993): 17–20, 34–41.

34. Wildavsky, op. cit., p. 48.

35. See, for example, K. S. Shrader-Frechette, "Ethics and Energy," in Tom Regan, ed., *Earthbound* (Prospect Heights, Ill.: Waveland Press, 1990). p. 121.

36. Wildavsky, op. cit., p. 27.

37. This happened between late 1982 and early 1983 when I was at the University of

Florida. First a few, then twelve, then fifty, then about a thousand leaking drums of Department of Transportation waste were discovered at an until-then unknown site, and then at another site near Gainsville, Florida.

38. Newton and Dillingham, op. cit., pp. 145–46.

39. A. Pablo Iannone, *Philosophy as Diplomacy* (Amherst, N.Y.: Humanity Books, 1994).

40. Editorial Staff, "Environmentalists Win Halt in Strip-Mining Revisions." *Christian Science Monitor*, Monday, April 19, 1982, p. 2.

41. L. Mouant, "EPA Hopes Michigan Waste Cleanup Settlement Will Serve as U.S. Model," *Christian Science Monitor*, November 26, 1982, p. 14.

42. This lassitude was evident in the United States in the 1980s. See, for example, "Hazards at the EPA," *Christian Science Monitor*, Tuesday, March 16, 1982, p. 28.

43. *Community Investment Corporation Demonstration Act of 1992, U.S. Code*, vol. 22, secs. 5305ff. (1994). See also John T. Metzger, "Community Development: The Clinton Strategy," *Journal of Housing*, March/April 1993, p. 84; and Rep. Henry B. González, "Viewpoint," *Texas Banking* 82 (October 1993), p. 14.

44. For details on these programs, see essay 5, "In Pursuit of Equity: Ethics, Distressed Communities, and the Community Reinvestment Act," in this book.

45. Stephen Breyer (op. cit., pp. 59–81) has suggested that environmental problems and issues be partly addressed through a centralized administrative group dependent on the executive and charged with a rationalizing mission. This may help; but, like other approaches I criticized in this essay, it is too limited and legalistic to have the effects sought.

Dialogue

— I am of two minds about your essay. On the one hand, I find it too restricted to superficial aspects of policy making about ecological matters. On the other hand, especially toward the end, the essay seems to be taking the right turn.

— Could you explain yourself in more detail?

— Let me try, beginning by what I find disquieting about the essay. You tend to discuss the immediate concerns involved—for example, the management of resources in dealing with Superfund-designated sites—not the overall ecological concerns affecting every aspect of human life now and in the future.

— Please do not misunderstand me. The overall concerns you mention are presupposed all along the essay. However, they have to do largely, if not exclusively, with the aims of policy and decision making concerning ecological matters. The essay, by contrast, deals with the problem of how to address those concerns through sound—meaning both effective and morally acceptable—policy making.

— Why, however, focus on this or that problem or issue to the detriment of dealing with the whole picture?

— Because not all, or even most, policy and decision problems concerning ecological matters are the same or nearly the same. For example, one of the problems I described, that posed by endangered species in Argentina, is crucially posed by lack of human resources. This problem is not even closely connected to that posed by cattle ranchers' encroachment in the Brazilian rain forest, a problem, indeed, an issue, where greed plays a crucial role. And none of these problems is closely related to those posed by exaggerated presumptions of knowledge and effectiveness concerning ecological risk management in some U.S. legislation.

— They all, however, affect planet earth.

— True. But not necessarily as a whole or in the same way. Indeed, this is fortunate. For otherwise we would be faced with the impossible situation of being unable to deal with any ecology-related problem unless we simultaneously dealt with each and every one or, at least, a great majority of them.

— Still, your piecemeal approach runs the risk of missing crucial overall interconnections between the problems.

— Not really. For my approach does pay attention to these interconnections so long as there is evidence that they exist and are significant. What the approach does not do is simply to assume the existence of significant interconnections, even when there is no reasonable evidence to do so.

— Are you, then, rejecting deep ecology?

— Deep ecology is not just a view but a movement. Hence, it makes little, if any, sense to reject it. At best, one may not join it. Further, the movement is quite varied, which makes it difficult to judge it as a whole. Now, I'd be the first one to point out that some ecology-related problems—say, those posed by ozone layer depletion and by the greenhouse effect—are global and, without addressing them, solving other, local problems may turn out to be ineffectual. But I heed the warning of one of the original proponents of deep ecology, Arne Naess, in his *Ecology, Community, and Lifestyle* (pp. 39–40). In global ecological approaches, there is always a danger of incurring the excessive universalization or generalization of ecological concepts. This is why, especially in this book where I explore the extension of ecological concepts to social contexts, I must be particularly cautious about the matter.

— You may be cautious about that. Yet, perhaps you are not as sufficiently cautious about the ecological consequences of prevalent human actions and lifestyles as you should be. Why not look at the overall planetary situation today and point to the need for radical lifestyle changes, beginning with population reductions?

— Because I am also cautious about what the overall situation is. Take population. Even if planetary population reductions are needed, how and where should they be implemented? Consider the case of Argentina. As I said in the previous essay, its area is about one-third that of the United States. Yet, it has only about thirty-eight million inhabitants and a relatively low rate of population growth. Also, except for about six million inhabitants, the rest live in urban areas. Bueños Aires alone has a population of about eight to ten million. In other words, much of the country is uninhabited. What sense would it make to suggest that Argentina decrease its population? Maybe it should do the opposite; for example, by encouraging immigration as it did in the past. If done selectively, this might help meet some of Argentina's needs for specialized human resources, say, the need for ecologists previously described.

— Maybe. But this, by itself, will not help with planetary population pressures.

— Granted. Nor will regulation by itself. This is why I am interested in a variety of voluntary approaches.

— That is why I am of two minds about your essay. Your focus on voluntarism would seem to point in the direction of deep ecology.

— As I said before, too many approaches fall under the heading of deep ecology. My essay may very well be in harmony with some of these. But this is not my main concern. My approach can perhaps be best understood as an attempt at moving beyond shallow policy-making approaches without ending up in the other extreme: a mere discussion of global-ecology-informed ends considered in a political and economic vacuum. I think this is not merely imprudent and irrational, but also morally objectionable, because it is unrealistic. Instead, my approach focuses on the processes of settling issues about ecology-related ends as these relate to currently feasible social decision processes for attaining ends.

— But your approach may undermine the whole thing by making the ends dependent on feasible processes. What if necessary ecology-related ends cannot be attained through such processes?

— If—and this is a big if—the situation you describe were real, we would be doomed. For we could not, in effect, do what is necessary for life on earth, hence us, to survive and flourish. Indeed, the implication of the situation you describe would be hopelessness, hardly a ground for seeking improvements in our interactions with the biota.

— But, even if not necessary, certain ecology-related ends might be desirable, yet unattainable through currently feasible social decision processes. Your approach would simply rule out such ends. Why?

— Given your assumptions, the ends you mention would be utopian because they would be desirable but unattainable through currently feasible social processes. Granted, utopian thought may have a function in helping our imagination to test the limits of what is actually attainable. After all, many apparently utopian ends have turned out to be attainable and attained in human history. But if they turn out to be irretrievably utopian, their pursuit will be merely a serious distraction. For when it comes to environmental problems and issues, time is often of the essence and, in such a situation, pursuing unattainable ends will backfire. In chasing a dream, we would lose all chances of improving our interactions with the biota. Indeed, these might worsen.

— But couldn't one at least work toward changing the currently feasible social decision processes so that ends that were utopian earlier, later become attainable?

— One certainly can and sometimes should do this. But the time frame for improving our interactions with the biota and the seriousness of the situation in need of correction should be assessed before embarking on the process of reform you suggest. Otherwise, as I just indicated, your suggestion might land us in irretrievable failure.

— Will you at least grant me that a sound approach to environmental problems and issues calls for some attitudinal change, hence, some reform of the current social situation?

— Yes, but it does not call merely for that. Indeed, I believe current general attitudes are highly underrated, and a sound approach to environmental problems and issues does not primarily call for attitudinal change across the board. Even if a good many people's attitudes need change, today's human beings evidence a good modicum of attitudes adequate for dealing soundly with problems and issues involving ecological concerns.

— How so?

— This is the subject of the next two essays.

In Pursuit of Equity: Ethics, Distressed Communities, and the Community Reinvestment Act*

by A. Pablo Iannone and Mary Kay Garrow

DISCRIMINATION IN CREDIT AND THE COMMUNITY REINVESTMENT ACT

In 1977, the United States Congress enacted the Community Reinvestment Act (CRA) as one of a series of statutes aimed at dealing with the problems faced by minority and low- to moderate-income groups in obtaining credit, and stopping redlining, the practice whereby financial institutions refused to lend money in specific sections of cities.[1] These were very laudable aims for at least two reasons. First, it was only fair to the groups who suffered the said discrimination and whose deposits the banks chartered in their communities and, engaging in the said discriminatory practices, had no qualms receiving. Second, the act, if effective, would have helped curb or prevent a variety of highly undesirable consequences—from unemployment to crime—in the communities affected by redlining and related forms of discrimination and disinvestment, in neighboring communities, and, indeed, in society at large.

The regulated financial institutions were largely insured banks and insured institutions as defined in section 401 of the National Housing Act.[2] They did not include nonbank financial institutions such as insurance companies and mutual funds. The CRA empowered the appropriate regulatory agencies to delay or deny the approval of applications for mergers, acquisitions, the opening of new branches, or the ability to receive deposit insurance on the part of regulated institutions that had not complied with the act. In this essay, we ask the following questions: Has the CRA attained its purposes? Has it been soundly implemented? What shortcomings have been involved in this implementation? What should be done about these shortcomings and the CRA, given the current economic and social circumstances in the United States? We will defend five theses. First, the CRA has attained its purposes only to a limited extent. Second, it has not been soundly implemented. Third, the implementation shortcomings involve not only lax enforcement but also, more crucially, an exclusive focus on adversarial administrative or legal approaches. Fourth, these shortcomings should be addressed not just by improving enforcement, but also by focusing on nonconfrontational approaches to improving the CRA's effectiveness, including but not limited to such things as negotiation, bargaining, mediation, arbitration, and consensus building. Fifth, current developments evidence a greater emphasis on such approaches, hence providing evidence for their political feasibility.

ENFORCEMENT LASSITUDE

In theory, the Community Reinvestment Act was supposed to be enforced by federal regulatory agencies. In practice, until the late 1980s, it was enforced by local community organizations that challenged regulatory approval of applications submitted by noncomplying financial institutions. In the CRA's early days, these were mostly savings and loan institutions and other thrifts seeking to open new branch facilities. When the wave of bank mergers began, however, the challenges were significantly redirected toward banks, and their number increased sharply. Thirty-five challenges were filed in 1987 alone, and the number of challenges filed between 1987 and 1989 amounted to almost half the number of challenges filed since the CRA's enactment.[3]

However, the enforcement was weak. For the first twelve years of the CRA's existence, federal regulators mostly delayed, but did not deny, approval of applications because of a CRA challenge. Only eight out of a total of fifty thousand applications had been denied on CRA grounds since 1978.[4]

This situation seemed to change on February 15, 1989, when the Federal Reserve Board (FRB) denied the applications of Continental Bank Corporation and Continental Illinois Bankcorp, Inc. (Continental Bank Corporation's holding company), to acquire the outstanding voting shares of Grand Canyon State Bank of Scottsdale, Arizona.[5] The grounds given primarily focused on noncompliance with the CRA and listed the compliance failures that led to the denial.

This decision prompted a great deal of speculation about the actual, complete grounds that led to it.[6] Yet, here, we will not pursue this matter. Instead, our purpose is to make a practical point. The FRB's decision increased the legal power of community organizations concerned with the fair implementation of the CRA. For it created a precedent and gave specific content to the grounds that, together, could lead to denying a financial institution's application.[7]

This precedent-setting decision was soon followed by a joint statement issued by the federal financial supervisory agencies whereby these announced their plans to increase enforcement and more closely evaluate compliance with the CRA, and listed activities which would demonstrate lender compliance with the act.[8]

CAN BETTER ENFORCEMENT SECURE EFFECTIVENESS?

Despite the FRB's decision and the federal financial supervisory agencies' determination, the until-then lax enforcement of the CRA was not corrected in an effective manner. Arguably, this resulted in part from the conflicting agenda created by the Financial Institutions Reform, Recovery, and Enforcement Act passed in the same year, which required fiscal conservatism on the part of savings institutions, coupled with the recessionary chances in the U.S. economy.[9] In addition, the effectiveness of the act was not significantly enhanced because of the defensive stand taken by banks, which largely interpreted and responded to the Federal Reserve Board's decision and the Federal Reserve System's joint statement merely as a signal to gather more documentation and to have it better organized.[10]

Indeed, this interpretation was reinforced by Federal Reserve Board Chairman Alan Greenspan who, also in 1989, indicated that CRA compliance would be measured by effective programs already in place, not by commitments to establish such programs. The burden of proof was shifted from the regulators needing to show noncompliance to the institutions having to show that they were complying with the CRA, but little else was done.[11]

Since this approach did not lead to substantial attainment of the CRA's purposes, hearings about the CRA were held in Congress in 1992. During these hearings, reports from community groups and local governments as well as various studies, including one by the Federal Reserve, revealed a widespread persistence of the situation the act had been enacted to correct. For example, minority applicants with comparable incomes were at least twice as likely to be rejected for loan applications as nonminority applicants.[12] Even today, this situation has not substantially changed, which leads to our first thesis: The CRA has attained its purposes only to a limited extent.

The CRA's failure to attain its purposes substantially after fifteen years of existence was largely seen, at the time, as an enforcement failure. True, as the previous discussion indicates, the act had not been soundly enforced, and its enforcement still needed much improvement. Yet one should be cautious about assessments of the act's effectiveness that focus on enforcement and adversarial approaches.

There are at least three reasons for this. First, as the previous discussion illustrates, by themselves, such approaches always come into effect too late, after there has been enough time for harm to be done. Second, the focus on community enforcement, adversarial approaches, and even outright confrontation, can hardly be touted as successful during the CRA's years of existence. If it had been, we would not be talking about the CRA's dismal implementation record, and events such as the Los Angeles riots might have been prevented. In short, such approaches could hardly suffice to address the U.S. need for urban and rural community reinvestment that the CRA was intended to help address.

Third, suppose that compliance could have been generally established by the said emphasis on greater and better organized documentation. It in no way follows that the CRA would have thereby been sufficiently effective in attaining its purposes. On the contrary. Mere compliance—especially, as it was generally the case, begrudging compliance—was, and always is, unlikely to lead to those purposes beyond, at best, a mere minimum.

These are significant reasons to conclude that merely confrontational— say, court or administrative adversarial—approaches were, and still are, bound to be inadequate to make the CRA effective. But they were the focus of CRA implementation for a long while, which leads to our next two theses: The CRA has not been soundly implemented; and the implementation shortcomings involve not only lax enforcement but also, more crucially, an exclusive focus on adversarial administrative or legal approaches.

ETHICS, SOCIOPOLITICAL PHILOSOPHY, AND THE CRA

Ethics and sociopolitical philosophy can make some helpful contributions toward dealing with the problems posed by the CRA and with its implementation. These problems, however, do not primarily, if at all, concern the fairness of the CRA's aims or the overall desirable consequences of its sound implementation. As briefly argued at the outset, there are at least two reasons for this. First, the CRA's aims were fair to the groups who suffered discrimination in credit through redlining and related forms of discrimination and disinvestment and whose deposits the banks chartered in their communities and engaging in the said discriminatory practices had no qualms receiving. Second, the CRA's aims, if substantially attained, would involve curbing or preventing a variety of highly undesirable consequences—from unemployment to crime—in the communities affected by the said discrimination and disinvestment practices, in neighboring communities, and in society at large.

Ethical and sociopolitical problems about the CRA arise at a different level: They primarily concern its implementation, which, as already argued, has had significant shortcomings because of a tunnel-vision focus on enforcement and adversarial approaches. Here, ethics and sociopolitical philosophy can make a contribution along the lines of pragmatic approaches and the conception of philosophy as diplomacy, which has been discussed in previous essays.[13]

Rather than expanding on these in a general manner, however, we will focus on the CRA and ask the following specific question: Are there additional, nonconfrontational, social decision-making procedures that did then, or can now, enhance the act's effectiveness?

NONCONFRONTATIONAL APPROACHES TO CRA EFFECTIVENESS

The limited effectiveness the CRA attained in the face of regulatory indifference, when not hostility, is of an extralegal and extra-administrative nature. The CRA gave negotiation and bargaining power to community organizations. For they learned "to use the challenge process to win commitments from banking institutions, usually in the form of negotiated settlements to the dispute."[14] These settlements were often very specific about what the financial institutions would do to address the credit needs of their neighborhoods. They often listed goals for credit products directed to these neighborhoods, such as mortgage loans, small business loans, and loans to nonprofit community development organizations. The settlements also listed such other goals as changes in loan underwriting and appraisal policies aimed at improving the availability of credit to low- and moderate-income groups; affirmative marketing activities targeting low-income and minority areas; funding for prepurchase housing counseling to make it more likely for first-time home buyers to obtain mortgages; and procedures for monitoring the attainment of these goals. Additional settlements included non-credit-related goals such as promises to provide low-cost checking accounts, to cash government benefit checks for nonaccount holders, to limit branch closings in certain areas, and to institute affirmative hiring policies.[15]

Some bankers responded to these settlements by charging that they were forms of credit allocation attained through blackmail.[16] This response, however, is too abstract to address the actual policy and decision problems that concern the CRA in a politically effective and morally sound manner. For it disregards the actual conflicts, mentioned at this essay's outset, that prompted the enactment of the CRA, the efforts aimed at enforcing it, and the negotiated attempts at dealing with the conflicts outside of the enforcement sphere.

A more defensible response has been to regard the negotiated settlements as quick, effective, and fair ways of resolving pending disputes.[17] Quick and fair they may be. It is not clear, however, that they are invariably effective. For the settlements are parasitic on challenges, hence on adversarial approaches to dealing with CRA noncompliance. Indeed, this (together with enforcement inconsistencies, examinations based on insufficient information, and the previously mentioned overemphasis on documentation to the detriment of attention to lending results) was one of the CRA enforcement shortcomings mentioned by the U.S. General Accounting Office in 1995.[18] The said parasitism on challenges, then, is a good reason for seeking still other, even less confrontational and more proactive approaches. This leads to our fourth thesis: The shortcomings in the past implementation of the CRA should be addressed not just by improving enforcement but also by focusing on nonconfrontational approaches to improving CRA's effectiveness, including, but not limited to, such things as negotiation, bargaining, mediation, arbitration, and consensus building.

One way of doing this is to form community reinvestment and development partnerships between banking institutions and local community development organizations. Some financial institutions and community organizations have had the wisdom to pursue this line of thought and action. Among the institutions that have been named as models for community development activities along these lines are South Shore Bank in Chicago, Illinois, and its Arkansas affiliate, Southern Development Bancorporation; the Center for Community Self Help in Durham, North Carolina; and Community Capital Bank in Brooklyn, New York. One of this essay's authors worked with People's Bank, whose headquarters are in Bridgeport, Connecticut, to develop a home improvement loan program that targets low- and moderate-income homeowners.

In addition, the U.S. Congress embraced community development banking in the housing reauthorization bill of 1992, which awards funds for the Community Investment Corporation Demonstration Act. For the years 1993 and 1994, it authorized annual sums of approximately $25 million in capital assistance through grants and loans, $15 million in development services and technical assistance, and $2 million to fund capacity-building training for the banks. Eligible banks can use the funds for direct lending purposes, Housing and Urban Development-approved equity investments, and mitigating the risk of other lenders through credit enhancement or partial funding of loan loss reserves.[19]

Many things remain to be worked out concerning the act's interpretation. For example, the characterization of an eligible bank is still unsettled. In addition, it has not yet been decided to which extent CRA ratings will be helped by participation in the activities just discussed.[20]

There are also at least two significant economic developments that may undermine the very notion of community development banking. First, the ever growing rate of bank mergers and consolidations has weakened the ties between the resulting huge banks and the communities at risk so much that, in general, these banks are likely to have become unable to serve the communities at all.[21]

Second, for quite a while, banks have been declining in importance relative to such nonbanks as mutual funds and finance companies. For example, in mid-1993, Fidelity Investments, the largest mutual fund company in the United States, had $200 billion in shareholder funds. This was about $50 billion more than Citicorp, the then-largest bank in the United States, had in deposits.[22] Since then, financial institutions other than banks have continued to seize a growing share of the market. As a result, there have been calls to make nonbanks comply with the CRA just as banks are required to do.[23]

Nonbank representatives have been quick to respond that, in contrast to banks, they do not make loans, do not have FDIC protection, and do not operate community branches. Yet, they increasingly have banklike features. Consider Fidelity Investments. It operates a network of walk-in investor centers in urban areas across the United States. It offers many types of accounts on which checks can be written, including a money market account on which there is no minimum on the amount of the check. It offers no-annual-fee MasterCard and Visa credit cards. And, though its accounts do not have FDIC protection, they are protected by the Securities Investor Protection Corporation, an entity established by Congress in 1970. That is, nonbanks have some connections with local communities that are not unlike those of banks. These may be weak; but the previously discussed growing weakness of the ties between banks and local communities strengthens the analogy between banks and nonbanks.

Whatever the strength of this analogy, however, the decline of banks and the emergence of nonbanks in traditional bank markets contribute to create a gap between local communities and both banks and nonbanks. As a result, the same point raised about large banks arises about nonbanks: Their very size and weak ties with local communities are likely to make them generally unable to serve these communities in the manner sought by the CRA and related pieces of legislation.

PROMISING ROUTES FOR COMMUNITY DEVELOPMENT

The preceding discussion provides some reason for trying additional routes toward community development. Along these lines, there has been talk of developing a different kind of parallel banking system, a "system of community development banks,"[24] which would replace the CRA and be set up to fill the vacuum left by recent bank mergers and consolidations.

This alternative need not entirely replace the CRA but only supplement it. After all, there are still reasons of equity and effectiveness that call for banks' and nonbanks' financial support of community development projects. Some legal teeth are needed to ensure at least begrudging support on the part of reticent banks, and simply getting rid of the CRA would lead to an enormous social cost in deteriorated communities that can recover through community reinvestment. However, the notion of a system of community development banks points in the right direction. For it further expands on some of the actual, less confrontational models previously mentioned. Indeed, the success of these models confirms that the problems they are meant to address can be better addressed along the less confrontational and more diplomatic approaches this essay proposes.

These approaches are not new. The Community Preservation Corporation (CPC), for example, was founded in 1974. It is a private nonprofit mortgage lender specializing in the financing of low- and moderate-income housing. At present, it is sponsored by more than fifty banks and insurance companies in the New York metropolitan area. Between 1974 and 1993, the CPC topped $1 billion in financing. This resulted in the rehabilitation or construction of more than 34,500 units, including 3,800 units of rehabilitated housing in Harlem, 6,100 units in the south Bronx, and 7,500 units in Washington Heights and Inwood.[25]

The CPC has set up a process whereby access to capital for low-cost builders has been increased, routinized, and simplified by working with government to provide a one-stop application for construction and permanent financing, as well as technical assistance.[26] By thoroughly advising prospective applicants at the outset, the CPC has secured a very low default record.[27] There is no reason other than bureaucratic inertia why this process could not be reproduced elsewhere.

Approaches to community development similar to those exemplified by the CPC have been adopted elsewhere in recent years. One example is the Massachusetts Housing Investment Corporation (MHIC), which was established by private banks in collaboration with community leaders and began operations in 1990.[28] Together with the Massachusetts Housing Partnership (MHP), MHIC has initiated a program that originates both the construction and the permanent loan for qualifying projects, manages the loan until construction is completed, and then sells the permanent loan to the MHP fund. This frees funds to start the process with other applicants.[29]

By 1995, MHIC had raised $52 million from participating banks. Of this amount, $35 million (67 percent) had been committed to thirty-two project loans comprising a total of 1,030 units of affordable housing.[30]

The General Assembly of Connecticut passed Public Act 93–404, An Act Establishing a Community Economic Development Fund (CEDF), on June 8, 1993. It directs the secretary of the Office of Policy and Management to establish a community development program in public investment communities and targeted investment communities.[31] The purpose of this program is to strengthen the neighborhoods by maintaining or creating employment for neighborhood residents, generating tax revenues, and stemming

physical deterioration and the problems resulting from deterioration by providing access to credit, facilitation of financing for community development
activities, and technical assistance.[32]

The CEDF is now in place, implemented by a network of private affiliates whose goals are coordinated. Two of these affiliates are the CEDF Corporation—a stock, for-profit corporation that will participate in limited
partnerships, joint ventures, and other investment vehicles—and the CEDF
Foundation—a nonstock, nonprofit corporation using public and private
sector funding to promote such things as skill development, technical assistance, and public-private sector communication. The foundation is governed
by a board of directors, at least one-third of which must be persons of low or
moderate income residing in communities for which the CEDF is intended.

This short list is in no way exhaustive, but it offers a somewhat detailed
idea of the models being used along the lines of the approaches suggested in
this essay. It also provides grounds for inferring our fifth thesis: Current
developments evidence a greater emphasis on such approaches, hence providing evidence for their political feasibility.

The said approaches use models that contribute to bring together different institutions such as banks, neighborhoods, and government, strengthening their interdependent interests in a manner fruitful to each. In this
regard, they focus on, and contribute to, enhancing the social ecology of
these institutions, in the sense pursued in this book, and already characterized in some detail in essays 3 and 4, and studied in the branch of sociology
known as *ecology*—the study of the spacing of people and institutions and the
resulting interdependencies.[33] The effectiveness and fairness of the arrangements thus worked out is more likely to be greater than that of the traditional, adversarial approach built in to the CRA and the related acts the U.S.
Congress enacted in the late 1970s. For it will tend to prevent community
deterioration rather than simply to correct it—and, judging from past experience, to correct it very slowly and minimally—once it begins to happen.[34]

NOTES

*This essay is based on a paper presented at the National Conference on Finance Ethics held
at the University of Florida, Gainesville, January 26–28, 1995. Ms. Mary Kay Garrow is executive vice president at Connecticut Housing Investment Fund, Inc., 121 Tremont Street,
Hartford, Connecticut 06105; (203) 233–5165.

1. *Community Reinvestment Act, U.S. Code*, secs. 2901ff. See also Senator Alan Cranston's
remarks in "Hearing Before the Subcommittee on Banking, Housing and Urban Affairs—
United States Senate," 102d Cong., 2d sess. September 15, 1992, in *Current Status of the Community Reinvestment Act* (Washington, D.C.: U.S. Government Printing Office, 1992), p. 1.
Examples of other statutes aimed at working together with the Community Reinvestment Act

in dealing with the said matters are the *Equal Credit Opportunity Act of 1975, U.S. Code*, vol. 6, secs. 1691 ff. (1994), and the *Home Mortgage Disclosure Act of 1975, U.S. Code*, vol. 4, secs. 2801–10 (1988). For a detailed discussion of these acts and their implementation structure, see Paul H. Schieber and Dennis Replansky, *The Lender's Guide to Consumer Compliance and Anti-Discrimination Laws* (Chicago: Probus, 1991). For a discussion of the Community Reinvestment Act's implementation during its first twelve years of existence, see Ann B. Shlav, "Financing Community: Methods for Assessing Residential Credit Disparities, Market Barriers, and Institutional Reinvestment Performance in the Metropolis," *Journal of Urban Affairs* 11, no. 3 (1989), pp. 201–23.

2. The National Housing Act required each appropriate federal financial supervisory agency to use its authority when examining regulated financial institutions and to encourage such institutions to meet the credit needs of the local communities in which they were chartered—including low- and moderate-income neighborhoods and minority neighborhoods—in a manner that was consistent with the safe and sound operation of such institutions. There were four appropriate agencies. First, the Office of the Comptroller of the Currency with respect to national banks. Second, the Board of Governors of the Federal Reserve System regarding state-chartered banks that were members of the Federal Reserve System and bank holding companies. Third, the Federal Deposit Insurance Corporation with respect to state-chartered banks that were not members of the Federal Reserve System and whose deposits were insured by the corporation. Fourth, the Federal Home Loan Bank Board regarding institutions whose deposits were insured by the Federal Savings and Loan Insurance Corporation, and savings and loan holding companies. See *Community Reinvestment Act*, op. cit., p. 1344. See also Federal Financial Institutions Examination Council, *A Citizens Guide to CRA* (Washington, D.C.: Federal Financial Institutions Examination Council, 1985), and *U.S. Code*, sec. 1724, pp. 911–13.

3. Allen J. Fishbein, "Banks Giving Credit with a Conscience," *Business and Society Review* (winter 1989): 34.

4. Ibid. See also Shlav, op. cit., p. 203.

5. Federal Reserve Board, "Order Denying Acquisition of a Bank," February 15, 1989, in Schieber and Replansky, op. cit., pp. 211–17.

6. According to one hypothesis, the decision was meant as a signal to financial institutions that they should take the CRA more seriously (Shlav, loc. cit.). By itself, however, this hypothesis failed to explain the change in the FRB's practice concerning the CRA's enforcement.

According to another hypothesis, the decision was a political maneuver to avoid strengthening CRA legislation (Federal Reserve Board, op. cit. See also Shlav, loc. cit.). Like all conspiracy hypotheses, however, this one was hard to support with evidence.

A third hypothesis was that the FRB had acted in response to a protest filed by the Amalgamated Clothing and Textile Worker's Union. The protest was based on two grounds. First, by giving credit to Japanese manufacturers, the bank's decision had led to the loss of jobs for union members. Second, Continental was using Federal Deposit Insurance Corporation (FDIC) bailout funds to remove itself from the retail banking market, thus eliminating services crucial to union members (Schieber and Replansky, op. cit., pp. 26–27). By itself, however, a challenge of this sort did not explain the change in the FRB's practice concerning the CRA's enforcement. After all, such challenges had been filed before and, at most, had led only to delays, never to application denials. In addition, the FRB's decision mentioned the matter only in passing (ibid.).

There was some evidence that the decision had been partly motivated by dissatisfaction with the FDIC's continued control of a great deal of Continental Illinois stock as a result of its earlier bailout of this bank (Editorial Staff, "Angell Persuaded Fed Board to Reject Continental's Bid; FDIC Ownership of Bank Cited as Primary Factor," *American Banker*, February 23, 1989, p. 1). Against this hypothesis, it has been argued that the fact that "the application rejection was grounded almost entirely on the failure of Continental Illinois to comply with its CRA requirements" (Schieber and Replansky, loc. cit.) evidences that federal agencies had become more willing to evaluate CRA compliance independently of any outside factors (ibid., p. 26).

But this is an exaggeration. For, first, however minimally, the FRB's decision mentioned such outside factors as the union protest and discontent with the FDIC's continued control of Continental's stock. And, second, other outside factors present at the time—such as recessionary trends in the U.S. economy and growing foreign competition for markets—had not been present before in the same mix and were highly unlikely to have no role whatsoever in a decision by the FRB.

A better hypothesis is that a combination of factors, including some of the outside factors just discussed, led to the FRB's decision and, as attested in the joint statement issued by the federal enforcement agencies, increased these agencies' willingness to strengthen enforcement and more closely evaluate compliance with the CRA.

7. "Order Denying Acquisition of a Bank," *Federal Register* 54, p. 13742, April 5, 1989. See also Schieber and Replansky, op. cit., pp. 28–30.

8. The statement included the following:

1. The bank did not monitor its CRA activities.

2. The bank did not promote its CRA activities.

3. The bank did not have a plan as to how it was to meet its CRA responsibilities.

4. The bank's CRA officer did not "interact substantially" with other bank staff.

5. The bank's board of directors had limited CRA discussion to mandatory annual approval of its formal CRA statement.

6. CRA statements listed products no longer routinely offered by the bank.

7. The bank made no significant effort to ascertain community credit needs.

8. The bank's level of participation in local community development efforts was unsatisfactory.

Schieber and Replansky, op. cit., p. 27, and Federal Reserve Board, loc. cit. See Federal Reserve System, Department of the Treasury, Office of the Comptroller of the Currency, Federal Home Loan Bank Board, and Federal Deposit Insurance Corporation, "Statement of the Federal Financial Supervisory Agencies Regarding the Community Reinvestment Act," *Federal Register* 54, no. 64, April 5, 1989, pp. 13742–46. For a discussion of this statement, see Schieber and Replansky, op. cit., pp. 28–33.

9. *Financial Institutions Reform, Recovery, and Enforcement Act of 1989, U.S. Code.* secs. 1790ff. See also Beth Linnen, ed., "Housing Equity Funds Test the Tolerance of FIRREA Rules," *Savings Institutions* (June 1990), pp. 12–14. In addition. a brief discussion of the nondecisive role a bank's CRA rating played in the acceptance or rejection of the bank's community support statements required by the Financial Institutions Reform, Recovery, and Enforcement Act can be found in Philip Britt. "FHFB Will Aim to Boost Community Lending Activity," *Savings Institutions* (July 1992), p. 9.

10. See Joseph Harrington, ed., "Credit Policies and Outreach Efforts Get Rigorous CRA Review," *Savings Institutions* (October 1990), pp. 9, 11. See also John C. Foreman and J.

McDuffie Brunson, "CRA: Bankers on the Defensive-Bankers Focus on Documentation of Community Lending Efforts in Response to New CRA Rules," *Bank Management* (January 1991), pp. 34–37, 40.

11. Foreman and McDuffie Brunson, op. cit., p. 36.

12. Senator Alan Cranston's opening statement, "Hearing Before the Subcommittee on Housing and Urban Affairs," op. cit., p. 2.

13. As indicated in previous essays and elsewhere, *diplomacy* here does not simply mean, as cynics would have it, the activity of doing and saying the nastiest things in the nicest ways. It means, quite broadly, the activity of dealing with states, nations, other social groups—such as the business sector, the financial community, and the U.S. urban and rural communities—and even individuals so that ill does not prevail (A. Pablo Iannone, op. cit., pp. 71–73). As for philosophy as diplomacy, it means a branch of inquiry aimed at dealing with a variety of problems—such as policy and decision problems—in ways that are feasible, effective, and crucially sensitive both to the often unsettled and conflictive nature of the concerns that contribute to pose the problems and to the variety of open-ended social decision procedures that may help settle these concerns and often deal with the problems through policies and decisions, and on the basis of reasons worked out in the policy-making process (ibid., pp. 75–76).

As also stated in previous essays, to say that philosophy as diplomacy aims at avoiding ill might lead one to think that philosophy as diplomacy is primarily a consequentialist notion—that is, that it takes the justifiability of policies and decisions to depend only on the value of their consequences. But it does not. For ill may consist in the violation of a right or failure to act in accordance with a duty or with principles of justice. And these are deontological considerations—that is, they take the justifiability of policies and decisions to depend only on their accordance with such things as rights, duties, and principles of justice (ibid., p. 73).

Nor is the conception of philosophy as diplomacy primarily deontological. It does not rule out the possibility that in certain cases—arguably in cases of widespread community deterioration—the social consequences are so catastrophic as to take precedence. And this need not be so because other deontological considerations take precedence. For the situations envisioned approach state-of-nature situations. And in any such situation, it is at least questionable whether deontological considerations such as rights or the obligations correlated with them carry much, if any, weight (ibid., p. 74). At any rate, the relative weight of these various considerations is at issue in such situations, and its determination needs to be worked out in the policy-making process. It is in this regard that social decision procedures take center stage (ibid., pp. 73–84).

14. Allen J. Fishbein, loc. cit.

15. Ibid.

16. Ibid. One might claim that the bankers' charge pointed to the fact that the CRA was yet another instance of reverse discrimination. This claim, however, is unjustified for at least two reasons. First, no reverse situation was created (as when minority members and women—some argue unjustly—get jobs that, were it not for affirmative action programs, would go to members of other groups). Second, whether or not an act institutes reverse discrimination hinges on the grounds for introducing the act and, in the CRA case, these grounds were stopping then-current discrimination and preventing community deterioration not just for the sake of the affected communities or their members but to prevent the highly undesirable, possibly catastrophic consequences from community deterioration for everyone or nearly everyone in the United States.

17. Ibid., pp. 34, 36.

18. In 1995, the U.S. General Accounting Office released a document assessing recent CRA implementation improvements and suggesting that, as stated in this essay, they may not

suffice to meet the CRA's goals. See U.S. General Accounting Office, "Community Reinvestment Act: Challenges Remain to Successfully Implement CRA," GAO/GGD-96-23, (Gaithersburg, Md.: GAO, 1995). For a commentary on this document, see "GAO Says Revised Regulations May Not Resolve CRA Problems," *Housing and Development Reporter* (December 4, 1995), p. 457.

19. *Community Investment Corporation Demonstration Act of 1992, U.S. Code*, vol. 22, secs. 5305ff. (1994). See also John T. Metzger, "Community Development: The Clinton Strategy," *Journal of Housing* (March/April 1993), p. 84; and Rep. Henry B. González, "Viewpoint," *Texas Banking* 82 (October 1993), p. 14.

20. Metzger, op. cit.

21. Associated Press, "Banking Industry Rebounds; Consolidation Continues," *New Haven Register*, May 17, 1994, p. D2.

22. Metzger, loc. cit.

23. Paul Starobin, "Make 'Em Pay," *National Journal*, July 24, 1993, p. 1856; Richard M. Rosenberg, "Viewpoint," *Texas Banking* (October 1993), p. 14.

24. Metzger, loc. cit.

25. The Community Preservation Corporation, *1993 Annual Report*, p. 13.

26. Ibid., p. 20.

27. Ibid., p. 13.

28. Massachusetts Housing Investment Corporation, *1993 Annual Report*, pp. 1–2.

29. Ibid., p. 1.

30. Ibid., p. 3.

31. ENT by CT; Department of Banking, September 14, 1993; 13:55, p. 1.

32. Ibid.

33. See, for example, Jess Stein et al., "ecology," entry no. 2, in *The Random House Dictionary of the English Language* (New York: Random House, 1970), p. 452. See also the preceding essays, especially essays 3 and 4, for more detailed discussions of the concept of ecology extended to social and related contexts.

34. See U.S. General Accounting Office, op. cit. and "GAO Says Revised Regulations May Not Resolve CRA Problems," op. cit. As for some recent, positive attempts at community development along the lines suggested in this essay, see Carolene Langie, "Recycling Black Dollars Is Working in L.A.," *New Haven Register*, Tuesday, November 19, 1996, p. D18.

Dialogue

— I am a bit frustrated by the preceding essay. I realize that, in the preface, you claimed that each essay can be read and studied as a more or less self-contained unit. The problem with the preceding essay is that it is too self-contained. I can see how social ecologies may be relevant to community development. But little work is done in the essay to show this in detail, and none is done to show the connection with philosophical ecologies.

— True. Much more needs to be done to show what these connections are. I will take up this topic in the next essay and proceed detailing it further in later essays. Doing this in the preceding essay would have been too superficial to be of help or too long to keep the essay's direction from becoming vague or lost.

— But at least you could tell us what, concerning social and related ecologies, is instructive about the nonadversarial models you present.

— What is instructive is that they point to the generally ignored, yet crucial, settled network of interactions and mutual adaptations between business, government, and the nonprofit sector in the United States and other societies.

— What do you mean?

— You will best understand what I mean by considering some of the questions that arise about the interconnections among these sectors. Here they are: What policy-making approaches are ethically and practically justified when working out the respective roles of government, business, and non-profits in providing for crucial social goods? How do existing laws and regulations—such as tax laws and labor-law constraints on voluntarism—affect the ability of nonprofits to provide for crucial social goods? How do different funding sources affect the ability of nonprofits to provide for these goods? What nonprofit approaches to matters of governmental policy, public attitudes, and business practices are practically and morally sounder in providing for social goods? For example, when, in what regards, and to what extent is advocacy advisable? How does the manner in which nonprofits operate and the ways in which governmental policy delineates how nonprofits relate to business and government affect democratic values and public participation in society? These are all questions that point to social ecologies that involve business, government, and nonprofits and the values, attitudes, practices, customs, and ideas operant in the interactions between business, government, and nonprofits.

— I can understand the questions and, perhaps, their relevance to understanding social and related ecologies. But I would prefer you tell me about this relevance.

— Let us begin by the preceding essay. It examined the limitations of legalistic approaches and argued for nonadversarial ways of addressing problems raised by redlining and related forms of discrimination and disinvestment in distressed communities. This is an area of human activity in which the network of interactions and mutual adaptations between business, government, and nonprofits is crucial and, in effect, partly constitutes a social ecology in the midst of which the activities occur and must be addressed.

— I can see now how the preceding essay discusses a type of case that can serve as a basis for extending ecological concepts beyond your previous discussion of multiculturalism. But how far can the extension go on this basis? After all, your case appears somewhat narrow when considered from the standpoint of economic relations generally.

— Granted, the case discussed in the preceding essay provides only partial grounds for extending ecological concepts to economic relations. The next essay, however, will provide further grounds, and later essays will argue analogous extensions to the art culture and the science and technology culture.

— Let us then move on.

— I will happily follow your suggestion. Move on we will.

Conscience at Work: Ethics, Technology, and Business Policy and Decision*

A World of Difficult Decisions

Technological developments do not take place in a vacuum. They are typically affected by the pressures of economic interests, political concerns, legal constraints, and moral considerations. In dealing with technology-related decisions, one should accordingly ask, What decisions should be made, by whom, and in what manner, *given these pressures*? Is, for example, moral heroism ever a moral requirement in making these decisions? Should it be? Can regulatory legislation play a role in giving individuals fair choices? Is it sufficient or even crucial for this purpose? What else can help?

With these questions in mind, I will examine and indicate some implications from two cases and their business environments. These implications will amount to six theses: First, as things stand, heroism is sometimes, if not often, needed in making the right decisions in technology-business. Second, however, heroism should not be necessary. Third, legislation can help give people in business a fair shake in dealing with difficult decisions. Yet, fourth, regulatory legislation, with its exclusive focus on penalties and adversarial approaches, is unlikely to be very helpful. Fifth, nonconfrontational approaches to social decision making, including, but not limited to, such things as negotiation, bargaining, mediation, arbitration, and consensus building can help and, indeed, are crucial in dealing with difficult decisions in business, including technology-business. Sixth, in these approaches, existing moral practices can and do play a role in the social ecology of business and society, which is largely compatible with technology-business considerations, in particular, and business considerations in general.

Let us next turn to the cases. The first concerns employees' moral dissent about technology-business decisions.

THE PINTO CASE[1]

On July 6, 1970, the first Pinto prototype was test-crashed at Ford. The windshield failed to behave in accordance with windshield-retention standards. In the tests made between July 6, 1970, and late 1971, this failure occurred again many times. Such a high number of test failures made it legally impossible to certify that the vehicle had passed the windshield-retention test (Federal Motor Vehicle Safety Standard no. 212).

Ford wanted the Pinto certified to comply with federal standards without delay, in order to meet the challenge of Volkswagen and the then-growing Japanese imports. Also, management had decided that, to be competitive, the Pinto had to weigh no more than two thousand pounds and cost no more than $2,000. This left few engineering options open to deal with the windshield-retention failures. To complicate matters further, the engineering department had already tooled for production before the certification testing began.

The chosen solution was to channel some of the kinetic energy generated in the crash away from the windshield, transmitting it via the driveshaft to the differential housing. This caused contact with the gas tank. But, at the time, windshield retention was, while fuel system integrity was not, a federally mandated area of certification. This fuel system weakness was known at Ford, but potential dangers were disregarded until years later, when Pinto tanks began exploding on impact in traffic collisions.

Frank Camps, the principal design engineer of the Pinto, felt uneasy about the cars' safety. He became even more uneasy when, in order to certify the Pinto at all costs, he was instructed to forget about failures and report only successes. But the successes were not forthcoming. So, Camps was further instructed to conduct the tests without dummies. With no dummies aboard, a car weight kept to its minimum, reports of only successes, and innumerable tests, the Pinto eventually passed the certification test in October 1971. Out of seven vehicles tested on a Saturday night, two measured up to federal standards and were submitted as federal certification cars. The other five were swept under the rug as developmental tests.

Camps made his concerns known in conversations with immediate supervisors. Yet, he took no further action for two years, as he said, for fear of losing his job. He had three children, a wife suffering from multiple sclerosis, and a mortgage to pay. After many attempts at resolving the matter with upper management through letter-writing and various meetings, and finding himself excluded from Ford's upper-level decisions concerning federal certification, Camps resigned in 1978.

BETWEEN HEROISM AND DUTY

Camps faced a *moral problem of conduct*: What should he do? Push aside his fears about the Pinto's safety, exhaust the corporate channels available to make his fears known, blow the whistle through the media or in some other way, sue Ford, resign, or some combination of these?

Given his moral obligations regarding his wife and children and the likelihood that his job would have been in jeopardy as a consequence, it is not clear that blowing the whistle would have been morally commendable. But, even if commendable, it arguably would not have been morally required but simply admirable as a heroic action.

One might be inclined to respond that moral heroism often is—as in whistle-blowing cases—morally obligatory in technology and business decision making. Yet, there are at least two reasons against this view. First, heroic actions are meritorious and so morally commendable because, characteristically, they involve going beyond the call of duty. If moral heroism became morally obligatory, the said reason for its being meritorious and morally commendable would disappear. One would simply have done one's duty. Second, treating moral heroism as morally obligatory is likely to undermine morality by making it hardly attainable. As the English economist R. H. Tawney once said, "The practical result of sentimentality is too often a violent reaction toward the baser kinds of Realpolitik."[2] Hence, though well-meaning, this exclusive reliance on outstanding character and the outmost virtues of moral leadership is likely to backfire. The preceding discussion provides reasons for concluding our first two theses: First, as things stand, heroism is sometimes, if not often, needed in making the right decisions in technology-business. However, heroism should not be necessary.

EFFECTIVENESS AND LIMITS OF LEGALISTIC RESPONSES

The type of situation Camps faced puts employees in extremely difficult positions. They get no fair shake. The difficulty was greater at Camps's time, because whistle-blowers had no legal protection. As a result, the understandable question arose: Should whistle-blowers be legally protected? After all, when individuals are placed in positions such as those described, and the invisible hand of the market obviously does not help, it is plausible that the difficulties may be eliminated by relying on the hand of government.[3]

There have been attempts at addressing the problem along the governmental lines just suggested. No law compels an individual to step forward and communicate any suspicions regarding criminal activity. But public policy favors exposing crime.[4] In the United States, federal legislation covering many areas of business and industry contains whistle-blowing statutes. Those including whistle-blowing provisions are the Air Pollution and Control Act; Comprehensive Environmental Response, Compensation, and Liability Act; Consumer Protection Act; Energy Reorganization Act; Longshoremen's and Harborworker's Act; Mine Safety and Health Act; Occupational Safety and Health Act; Railroad Safety Act; Safe Drinking Water Act; Selective Service Act; Solid Waste Disposal Act; Surface Transportation Assistance Act of 1982; Toxic Substances Control Act; Vocational Rehabilitation Act; and the Water Pollution Control Act.[5]

Many states have also addressed the problem as it arises in the public sector; but resulting state laws vary widely in coverage. Some apply only to public employees, others apply to public employees and employees of public contractors, and still others apply both to public employees and employees in the private sector.[6] And while some people consider even the strictest of these laws and the interpretations accompanying them too weak, others think the weakest are too strong. In any case, the enactment of whistle-blowing legislation has provided some protection for whistle-blowers without exposing firms to unfair charges of wrongdoing. Hence, our third thesis: Legislation can help give people in business a fair shake in dealing with difficult decisions. But a further question remains: Is regulatory legislation, however improved, likely to be sufficiently effective in soundly dealing with the public policy matters of concern?

There is good reason to think that it is not. For, at best, like all adversarial approaches, which typically focus on enforcement and punitive responses, regulatory legislation at best leads firms to *reluctant* obedience. In fact, it often leads largely to the production of record-keeping piles of paper and contributes to the well-known court case overload in the United States.[7] Our fourth thesis follows: Regulatory legislation, with its exclusive focus on penalties and adversarial approaches, is unlikely to be very helpful.

This is not to say that the hand of government is not needed to deal soundly with moral problems posed by employee moral dissent and, for that matter, with other problems posed by conflicts between technology-business firms and society. It is simply to say that exclusive reliance on the hand of government is not enough, has significant undesirable consequences, and, indeed, often backfires.

Toward Nonadversarial Approaches

Just as in cases discussed in previous essays, a tunnel-vision focus on adversarial approaches undermines the policy matters of concern involved, not only in cases of whistle blowing in technology-businesses, but also in cases involving technology-business policy and decisions generally. As in the cases discussed in previous essays, ethics and sociopolitical philosophy can here make a contribution along the lines of pragmatic approaches and the conception of philosophy as diplomacy, which I have developed elsewhere and already briefly discussed in this book.[8]

Along these lines, nongovernmental approaches have been suggested and effectively used. They range from improving professional codes of conduct and technology-business firms' internal policies to creating a more proactive organization of these firms and their relations with society and to increasing the firms' reliance on social decision procedures such as negotiation, bargaining, mediation, and arbitration.[9] In general, these attempts shift the focus of reliance from the traditional policy instruments, the invisible hand of the market and the hand of government, to the hand of management.[10]

This makes them crucial because, if we followed what, according to received—and not quite historically accurate—opinion, was Adam Smith's advice, and merely relied on the invisible hand of the market, things would take too long to straighten out—if ever.[11] As Lord Keynes is reported to have said: "In the long run we are all dead." On the other hand, if we merely relied on the hand of government, things would often fail to straighten out. For, as Adam Smith correctly saw, governmental regulations are often ineffective and, when governments are too closely involved with business, they develop too much of an inclination to play favorites.

At any rate, Adam Smith's reputed suggestion never became a reality. Contemporary governments regulate businesses in innumerable ways. And while many business firms often try to safeguard their interests when faced with regulation, one can also frequently find business firms trying to win governmental favors. No doubt, a good number of business firms behave reasonably well. But there is a widespread belief that their good behavior is the result of regulation alone.[12]

This is why the third suggested approach, which would rely on the hand of management, is often met with cynicism. People ask, Why expect such measures to be anything but mere public relations efforts? The evidence is, however, that technology-business policies, decisions, and organizational efforts sometimes go well beyond the level of mere public relations. Let us consider our second case.

MERCK AND RIVER BLINDNESS

The antiparasitic drug Mectizan was developed by Merck and Co., Inc. It is effective in curing, with only mild side effects, onchocerciasis, or river blindness. People contract river blindness by walking barefoot in tropical areas, where the wet ground is home to the germs that cause the disease. As a result, they go blind.

After the French Directorate of Pharmacy and Drugs approved Mectizan, Merck did not market it. Instead, in October 1987, it offered it, free of charge, to developing countries where the disease is found. The only condition was that, before delivering the drug, a review committee formed by the company in consultation with the World Health Organization would have to certify distribution plans from countries applying for free supplies.[13]

One could try to figure out some nonmoral motivation for Merck's actions—say, public image. Yet, it is doubtful that, from a strictly business standpoint, unremarkable public image gains will outweigh development and production costs. After all, Merck's free distribution of Mectizan has hardly been widely known across the planet, and its effect on sales is unlikely to have been, or ever become, significant even among consumers knowledgeable of Merck's action. A better explanation is that Merck did what it did because its officials had moral reasons to do so.

One might object that shareholders were shortchanged by incurring development and production costs that led to no returns. But this line of inquiry is irrelevant to the point at issue. For, first, there appears to have been no shareholder revolt at Merck as a result of the development, production, and free distribution of Mectizan. This evidences that the objection is based on an imagined situation, not on the actual thinking of those involved. Also, it suggests that moral reasons were operant not only among Merck's management but also among its shareholders.

Second, the point at issue is whether Merck used moral reasons in deciding to distribute Mectizan free of charge. But the fact that Merck should not wantonly shortchange its shareholders is a moral reason. So, even if the objection turned out to be well grounded, it would only prove that Merck did not use moral reasons well, not that it did not engage in moral reasoning at all.

Hence, instead of trying to pull imaginary objections out of thin air, it seems reasonable to assume the obvious: Moral reasons—which include but are not exhausted by Merck's moral obligations to its shareholders—were significantly operant in Merck's decision. There is not good reason for surprise here. After all, businesspeople are ordinary people who engage in business. Hence, as a group, and like any group of people, they include some of outstanding moral character, some with serious character faults, and a majority who have enough common moral sense to act, much of the time, substantially in accordance with moral practices.

THE MORAL EQUIVALENT OF THE OZONE LAYER

The preceding discussion makes plain that neither an exclusive reliance on outstanding character, nor an exclusive reliance on the law (or, for that matter, on the market), suffices for dealing with technology-business policy and decisions. A modicum of sound moral practices is also crucial.[14] The discussion also provides reasons to believe that a modicum of sound moral practices is in fact operant in business.

One might object, however, that the reasons so far given at best establish that there are some business cases in which moral reasons have been decisive, not that these cases amount to a modicum of sound moral practices. However, the cases this essay discusses are not mere flukes. People in business frequently and honestly use moral reasons in discussing business activities. For example, U.S. managers sometimes argue that the dumping of foreign products—say, of towels by Brazilian companies—in the United States is an unfair business practice that harms U.S. firms by driving them out of business, thus also harming their owners and employees and, in the long run, U.S. consumers, whose market choices come to be made abroad. The same type of argument has been heard for many decades in developing countries concerned with defending their incipient national industry. Regardless of whether these moral reasons are good, they evidence that moral reasons are widely used in business.

One might respond that such reasons are merely a front for selfish motives. Granted, moral reasons are sometimes used as a front for immoral purposes; but there is no good reason for holding that, whenever moral reasons are given, they invariably amount merely to a front. On the contrary. For, as already argued concerning the Merck case, businesspeople are ordinary people who engage in business, hence, they include a variety of people, many of whom take moral reasons seriously and use them in their decision making.

Of course, one might insist that all or most individuals are selfish and use moral reasons as fronts. This view, however, cannot be based on evidence, for, as indicated, the evidence points in the opposite direction. In effect, though the view is ostensibly factual (namely about facts of human psychology), it can be defended only discounting as a rule all evidence that could disconfirm it. But this is to hold on to a position on faith, thereby leaving the realm of rational inquiry and critical scrutiny. Further, since the view is—and is claimed as—factual, yet it is defended by making it, in principle, impervious to disconfirming evidence, the position thus taken involves a contradiction: that the view is factual, yet (because unassailable by evidence) it is not factual.

One still might argue that, even if operant, moral reasons and sound moral practices invariably take a backseat in business, because the overriding nature of profitability and growth is bound invariably to relegate moral considerations to such a secondary position. Indeed, one might even argue that profitability and growth should always take precedence in business and that cases in which they do not only evidence the unreasonableness of those businesspeople who use moral considerations at all.

This view has been advanced by a variety of authors. Albert Carr, for example, wrote:

> In the prevailing concept of corporate efficiency, a continual lowering of costs relative to sales is cardinal. For low costs are a key not only to higher profits but to corporate maneuverability, to advantage in recruiting the best men, and to the ability to at least hold a share of a competitive market.[15]

And he later added:

> When the directors and managers of a corporation enter the boardroom to debate policy, they park their private conscience outside.
>
> If they did not subordinate their inner scruples to considerations of profitability and growth, they would fail in their responsibility to the company that pays them.[16]

A variety of objections have been raised against this view.[17] Here, however, I would like to focus on an aspect of this view that has not received much attention. Carr's position is based on a general, though by no means universal, conception of the purpose of business firms and, indeed, of the institution of business itself. This purpose is profitability and growth or, at least, a minimum of the market share necessary for profitability and growth.

No doubt, one crucial aim pursued by any business firm—including those operated at a loss in order to pay less taxes—is to bring profits, the more the merrier, to its owners or investors, and accordingly grow to attain this aim. But this is not to say that such aim is the purpose of business firms or of the institution of business. Indeed, though such a belief is general, it is by no means universal in business. For example, Douglas S. Sherwin, who for years headed a Phillips Petroleum subsidiary in Oklahoma, expressed a very different conception of the purpose of business firms and the institution of business in which they operate:

> Profit is the purpose that the owners have in risking their capital. It is the reward for bearing risk. And the owners' requirement for profit supplies one of the great disciplines that makes business an economic system. But profit is not the purpose of business.[18]

As to the nature of this economic system, he says:

> At its core, business is a feedback system. Capital owners, employees, and consumers are the members of the system. They coproduce its output. . . .
>
> What is ofttimes not appreciated, even by the members themselves, is that the system's members are *interdependent*. Together, and only together, can they produce the output they subsequently share.

Besides being interdependent, the members of the system are entirely equal in importance. Business people often claim primacy for capital, perceiving it as the fuel of enterprise, while consumers tend to assume that the whole point of business is to provide them with goods and services. But no members of a system can be primary. Since the contribution of every member is necessary and no contribution is sufficient, the members are equal.[19]

Sherwin's view of business is much more sophisticated than the simplistic views he criticizes. In it, there is not one but various purposes of business and business firms—from profits as rewards for risks taken by investors, through goods and services to satisfy consumer demand, to income-yielding jobs for employees. Further, as Sherwin was also quick to point out, these purposes are pursued through business activities rather than, say, through a directed economy, because public policy both makes it possible and circumscribes the scope of business activities, typically on the grounds that business is a much more effective way than directed economies to attain the said purposes.

Given that public policy both institutes and regulates business, one might want to argue, against Sherwin, that business is a feedback system whose components are not three, as he said, but at least four: investors, consumers, employees, and government. This, however, is a line of thought I will not pursue here. Instead, I will summarize five facts brought out by our discussion in this section. First, business activities take place within a variety of interactions—from negotiation and bargaining to mediation, arbitration, and consensus building—between investors, consumers, employees, and policymakers. Second, these interactions are aimed at working out ways of attaining interdependent social goods. Third, they involve a variety of moral practices. These range from those concerning honesty in negotiation and fairness in bargaining to moral practices partly formulated in the codes of ethics of various firms, business and industry associations, and other business and technology community organizations. Fourth, the fact that these practices are involved in actual business practice concerning technology-business and other forms of business evidences that they are largely compatible with technology-business considerations, in particular, and business considerations in general. Indeed, fifth, the network of all these practices and the interactions they involve arguably constitutes a *social ecology* of business and society in the sense of social ecology initially characterized in previous essays and further specified in the book's remainder.[20]

These facts involve our remaining theses: Fifth, nonconfrontational approaches to social decision making, including, but not limited to, such things as negotiation, bargaining, mediation, arbitration, and consensus building can help and, indeed, are crucial in dealing with difficult decisions in business, including technology-business. Sixth, in these approaches, existing moral practices can and do play a role in the social ecology of business and society, which is largely compatible with technology-business considerations, in particular, and business considerations in general.

Business policies and decisions must be evaluated and made within the framework of the social ecology just described in order for them to lead to feasible and effective arrangements. Indeed, evaluating and making such policies and decisions in disregard of the said ecology amounts to being irrelevant, because it is likely to lead to unfeasible and ineffective arrangements. It is fortunate that, as indicated, the said social ecology, which centrally includes moral practices and reasons, is substantially operant. For it is the moral equivalent of the ozone layer: Without it, we are gone.[21] It is, then, our ongoing task as engineers policymakers, or simply members of society, to help preserve and strengthen it.

NOTES

*This essay is based on a paper presented at the Ethics in the Workplace Panel Discussion held at the University of New Haven, November 16, 1994.

1. This account is significantly based on Frank Camps's own account, "To Design a Pinto," which appeared in *Business and Society Review*, "Is Whistle-Blowing the Same as Informing?" (fall 1981), pp. 215–18. Camps's selection was an excerpt from an article by Camps originally included in Alan F. Westin, *Whistle-Blowing!* (New York: McGraw-Hill, 1981).

2. R. H. Tawney, *Religion and the Rise of Capitalism* (New York: Mentor, 1958), p. 156.

3. I have discussed the various kinds of moral (or ethical) problems in my *Contemporary Moral Controversies in Technology* (New York and London: Oxford University Press, 1987), pp. 4–5, and *Contemporary Moral Controversies in Business* (New York and London: Oxford University Press, 1989), pp. 4–5. For a discussion of the invisible hand of the market as compared with the hand of government and the hand of management, see Kenneth E. Goodpaster and John B. Matthews Jr., "Can a Corporation Have a Conscience?" *Harvard Business Review* (January–February 1982), pp. 132–41.

4. *American Jurisprudence*, vol. 82, 2d ed., rev. (Rochester, N.Y.: Lawyers Cooperative Publishing, 1992), pp. 729, 731.

5. Ibid., pp. 731–76.

6. Public Law 101–12 [Senate Bill no. 20], April 10, 1989.

7. For a discussion of these matters, see my *Philosophy as Diplomacy* (Amherst, N.Y.: Humanity Books, 1994), pp. 3–4.

8. See, for example, essays 4 and 5 in this book, and my *Philosophy as Diplomacy*, passim.

9. See, for example, Douglas S. Sherwin, "The Ethical Roots of the Business System," *Harvard Business Review* (November–December 1983), pp. 183–87, 189–92, also found in my *Contemporary Moral Controversies in Business*, pp. 35–43; James A. Waters, "Catch 20.5: Corporate Morality as an Organizational Phenomenon," *Organizational Dynamics* (spring 1978), pp. 3–19, also in *Contemporary Moral Controversies in Business*, pp. 151–63; and *Philosophy as Diplomacy*, passim.

10. See Goodpaster and Matthews, op. cit.

11. For an enlightening discussion of Adam Smith's actual position, see Patricia H. Werhane, *Adam Smith and His Legacy for Modern Capitalism* (New York and Oxford: Oxford University Press, 1991), passim, especially pp. 4–5.

12. For a brief but instructive discussion of these points, see David Braybrooke, *Ethics in the World of Business* (Totowa, N.J.: Rowman & Allanheld, 1983), pp. 405–406. See also my *Contemporary Moral Controversies in Business*, pp. 98–104 and the selections discussed in those pages.

13. I have discussed this case in *Contemporary Moral Controversies in Business*. A more detailed report can be found in John Walsh, "Merck Donates Drug for River Blindness," *Science* 238 (October 30, 1987): 610. A case worth discussing in this connection is the Tylenol case, a detailed account of which appeared in Rogene A. Buchholz, *Fundamental Concepts and Problems in Business Ethics* (Englewood Cliffs, N.J.: Prentice Hall, 1989), pp. 212–27.

14. David Brown, *Cooperation Among Strangers* (Dayton, Ohio: Kettering Foundation, 1992), p. 5.

15. Albert Carr, "Can an Executive Afford a Conscience?" *Harvard Business Review* (July–August 1970), pp. 58–64, also in my *Contemporary Moral Controversies in Business*, pp. 23–29 (the quotation appears on p. 27).

16. Ibid., *Contemporary Moral Controversies in Business*, p. 28.

17. See, for example, Norman C. Gillespie, "The Business of Ethics," *University of Michigan Business Review* 26, no. 6 (November 1975): 14, also in my *Contemporary Moral Controversies in Business*, pp. 30–34.

18. Douglas S. Sherwin, "The Ethical Roots of the Business System," *Harvard Business Review* (November–December 1983): 183–87, 189–92, also in my *Contemporary Moral Controversies in Business*, pp. 35–43 (the quotation appears on p. 37).

19. Ibid., *Contemporary Moral Controversies in Business*, p. 36. For a characterization of the features of feedback systems, Sherwin refers the reader to C. West Churchman, *The Design of Inquiring Systems* (New York: Basic Books, 1971). The literature on this and related topics is quite substantial. See, for example, Karl J. Astrom, *Adaptive Control* (Reading, Mass.: Addison-Wesley, 1989), Ch. 1; Byrnes and A. Kurzhanski, eds., *Modelling and Adaptive Control* (Berlin and New York: Springer-Verlag, 1988); John H. Holland, *Adaptation in Natural and Artificial Systems: An Introductory Analysis with Applications to Biology, Control, and Artificial Intelligence* (Cambridge, Mass.: MIT Press, 1992); P. V. Kokotovic, ed., *Foundations of Adaptive Control* (Berlin and New York: Springer-Verlag, 1991); Otto Mayr, *Feedback Mechanisms* (Washington, D.C.: Smithsonian Institution Press, 1971); George P. Richardson, *Feedback Thought in Social Science and Systems Theory* (Philadelphia: University of Pennsylvania Press, 1991); Norbert Wiener, *The Human Use of Human Beings* (Garden City, N.Y.: Doubleday, 1954).

20. See, for example, Jess Stein et al., "ecology," entry no. 2, in *The Random House Dictionary of the English Language* (New York: Random House, 1970), p. 452. For discussions on social ecology understood along the lines of the characterization given in this essay's text, as focusing on the relationship of humans to their sociocultural environment, see Ramachandra Guha, ed., *Social Ecology* (Delhi and New York: Oxford University Press, 1994); Tim Hayward, *Ecological Thought: An Introduction* (Cambridge, England: Polity Press; Cambridge, Mass.: Blackwell, 1995); Martell Luke, *Ecology and Society: An Introduction* (Cambridge: Polity Press, 1994); Carolyn Merchant, ed., *Ecology* (Amherst, N.Y.: Humanity Books, 1994); Rudolf H. Moos and Paul M. Insel, *Issues in Social Ecology: Human Milieus* (Palo Alto, Calif.: National Press Books, 1974); Roy Morrison, *Ecological Democracy* (Boston: South End Press, 1995); Susan Leigh Star, ed., *Ecologies of Knowledge: Work and Politics in Science and Technology* (Albany: State University of New York Press, 1995).

21. For a discussion related to these points, see William Henry Hay, "Under the Blue Dome of the Heavens," presidential address delivered before the Seventy-third Annual Western Meeting of the American Philosophical Association in Chicago, April 25, 1975, *Proceedings and Addresses of the American Philosophical Association*, vol. 48 (Newark, Del.: APA. 1974–75), pp. 54–67.

Dialogue

— I thought I was beginning to understand the sense in which you use the expression *social ecologies*, but then you introduced the concept of feedback systems to explain business and its social ecology. Result: I have more questions about social ecologies than before. What is a feedback system? Are there types of feedback systems? If so, what type does business belong to? Also, are all social ecologies feedback systems? Are they all of the same type?

— Your questions are valid; but they need not undermine your understanding of social ecologies. Perhaps your puzzlement is prompted by the fact that my inquiry concerning social ecologies is twofold. On the one hand, I use social ecologies to clarify various networks of interaction and mutual adaptations in human life—such as those among various cultures or their members, those among humans and the natural ecologies in whose midst they flourish, and those between business and society. On the other hand, in order to strengthen the explanatory function of social ecologies, I try to characterize them in further detail than I have done so far. This is where the concept of feedback systems comes in.

— But you invoked this concept, following Sherwin, with regard to business. Is it applicable beyond that?

— Yes, but I will more fully explain how and to what extent in the remaining essays.

— So, all we have thus far is not an explanation but merely the promise of an explanation of the concept of social ecologies as feedback systems.

— We have more than that, namely, an explanation of one social ecology as a feedback system-that of business in its interactions and mutual adaptations with society.

— But you have not explained how you conceive of feedback systems.

— Not in general; but, with Sherwin's help, I have explained how I conceive of the feedback system constituted by business.

— So where do we stand regarding all this?

— First, we have an initial example, that of business, which begins to help clarify the concept of social ecologies as feedback systems. Second, the social ecology of business constitutes an additional case to be added to our previous list of social ecologies of interactions and adaptations among different cultures and their members. This strengthens the case for generalizing the social ecology model to still other areas of human activity.

— I am curious to see how you do this. Here, however, I would like to raise another set of questions: What are the normative implications of social ecologies? Do they have any? Or are you simply engaged in a merely descriptive inquiry?

— They do have normative implications. In fact I will begin to outline them in the next essay.

— Do you, then, believe that nature can tell us what is right and what is wrong?

— We need to be careful with the sense in which we use the term nature. I believe your question has a modestly affirmative answer so long as nature is understood as, the world—including the social world—in which we live and which is not entirely dependent on the beliefs, attitudes, or preferences of any one individual or group.

— I suppose that is one of the various senses — of the term nature. But what do you mean by saying my question has a modestly affirmative answer?

— I mean that the world in which we live, which is independent of any one of us and of any group, establishes a few, but crucial, limiting conditions concerning what actions or decisions are right; what policies, practices, customs, or institutions are justified; and what attitudes or character traits are good.

— What are these conditions?

— A central one is this: Whenever they involve clearly unrealistic expectations or aims whose pursuit would lead to significant harm—for example, to violations of actual rights or to undesirable consequences—for unwilling and innocent individuals or groups, then actions and decisions seeking such aims are wrong; proposed new policies, practices, customs, and institutions seeking such aims are unjustified; and envisioned attitudes and traits seeking such aims are objectionable or bad.

— Why should realism matter normatively?

— Because rationality implies realism. After all, one obvious way of being irrational is being out of touch with reality.

— But realism can be a prescription for paralysis.

— True. That is why I qualified my statement. Remember, the condition applies when, and only when, the expectations or aims involved are *clearly* unrealistic and would lead to significant harm for unwilling and innocent individuals or groups.

— Think of how much good this condition may rule out!

— Only the good that belongs with fantasy and constitutes no reliable option while likely to lead to significant harm for those who are innocent and unwilling to undergo it. You are beginning to forget that the implications I have in mind are quite modest. As I have argued at length in *Philosophy as Diplomacy*, all that the condition I indicated implies is that we should take seriously the available evidence about how likely or unlikely are the circumstances and consequences of actions, decisions, policies, and other matters of moral and political evaluation. We should not flippantly disregard their probability or improbability for the sake of chasing after utopias or wishful thinking. If such disregard is ever justified, the burden of proof is on us to show why.

— Even if this condition is valid, what does it have to do with social ecologies? I feel I lost your line of thought.

— Both natural and social ecologies, and their interconnections with human and nonhuman flourishing, serve as bases for constraints on what moral agents can do, institute, or bring about. In other words, entertained actions and decisions are presumed wrong; proposed policies, practices, customs, and institutions are presumed unjustified; and envisioned attitudes and traits are presumed bad or objectionable, whenever they are likely to destabilize social or natural ecologies in a significant manner and these ecologies are not ridden with oppression, exploitation, or parasitism to the core. If they are still morally acceptable, reasons need to be offered to show they are acceptable despite the said presumptions.

— What could these reasons be?

There is no clear-cut recipe because, as I argued in essay 4, some of these reasons are in the making. But the reasons might, for example, point out that there is a clear and present danger that the affected ecologies will collapse anyway, and the activities or arrangements considered would favor surging, and more reliable, new ecologies on which humans could flourish.

— This is an overly optimistic view. What if a merely biological ecology would be destabilized by human activities that would lead to generalized human flourishing?

— How frequent is such a situation?

— Frequent enough. The snail darter fish you mentioned in essay 4 is a case in point.

— I do not think it is. For, first, all economic studies done at the time indicated that human flourishing in the Tellico Dam area was not dependent on but, in fact, undermined by the building of the dam. Second, though at one time it was thought that the cave where the snail darter fish were initially discovered was the fish's only habitat, it later turned out that there were other caves with snail darter fish in the vicinity that were unaffected by the dam's construction. Third, the snail darter fish found in the original cave did very well after being transferred to a nearby cave unaffected by the dam's construction. And fourth, and foremost, the snail darter fish were a species, not an ecology—which is what I have been talking about.

— I can see your point. But I can mention ecosystems on which human flourishing does not depend at all. Some derive their energy purely from chemical sources, such as bacteria living underground (*Science News*, March 29, 1997, pp. 192–93) or deep-sea communities that feed off mineral-rich hydrothermal vents (*Science News*, September 7, 1996, pp. 156–57). Recently, forty-eight animal species—including thirty-three new ones—were found in Movile Cave, under a Romanian cornfield. They are part of a food chain that draws sustenance only from energy-rich molecules in rocks—in other words, without the use of photosynthesis (*Science News*, June 29, 1996, p. 405). These are ecosystems in which human flourishing could very well ram through.

— Here you have provided good evidence for the view I have suggested that not every ecology is connected to every other ecology. The ones you mentioned are largely isolated from the usual ecologies within which humans flourish. Hence, they are not likely to conflict with human flourishing in the manner in which you suggested. Indeed, now that they have been discovered, they may be crucial for human flourishing widely understood—as I believe it should—so as to include the growth of human knowledge.

— But what about ecologies not thus isolated?

— Let us put aside the special cases where all human flourishing depends on some such ecology—such as on the ecology of rain forests, which regulate weather patterns, hence are crucial for the existence of agriculture, hence of humans on earth. In the remainder of cases, one (at least typically) finds ecologies on which the flourishing of some, though not all or most, humans depends. The problem then becomes one of resolving a conflict between what is in the interest of different humans, not one that pits humans against some aspect of nature of no consequence to any humans.

— Even if you are correct, your position still seems overly optimistic. For you seem to assume that ecological change can be managed.

— I would not rule it out to some extent. Yet, no doubt, there is a serious danger of humans becoming arrogant about their ability to manage ecologies. They may become overly sure that they know what they are doing, or overly sure that they are in control of such changes.

— In order to put these things in perspective, it would be good to examine some examples of human activities as they have affected, perhaps intentionally, perhaps inadvertently, natural and social ecologies.

— You seem to be reading my mind. In the next essay, I will do precisely what you suggest.

— Good. I only hope that your attempt does not land me in even greater and deeper confusion than I suffer now.

— I hope so too. . . . I hope so too.

PART 3

Social Ecologies: Human Groups, Habitats, and Philosophy

GEOGRAPHIES OF CONFLICT

Leonardo da Vinci's posthumous writings tell us of a dream city-block. It has two levels. Above, there are no carts. It is a place for the enjoyment of those who can afford it. Beneath, sweaty workers and merchants, using carts and beasts, earn their livelihood by ensuring that the enjoyment and good life above can go on.[1] No doubt, this seminal conception of zoning in a city block reflected the economic stratification of da Vinci's time. More important, however, it reflected the fact that da Vinci, like so many of his Italian Renaissance contemporaries, was sedentary and urban while quite aware of the quiet of the country villas and, generally, nonurban life. As someone with a strong interest in life's enjoyments and the arts, he saw traffic, noise, dirt, and the sheer bustle of everyday business and work as a nuisance.

Contemporary traffic jams, noise, and pollution in much urban life on earth, together with the generalized objections against this situation—from those based on health risks to those based on aesthetic preferences—that are voiced around the planet, make it clear that da Vinci's concerns are still relevant. Also relevant, however, are the widespread objections—from those adducing economic discrimination to those accusing zoning bodies of elitism—voiced against the manner in which zoning has often been applied in trying to address the said health-related and aesthetic concerns. Equally relevant—and also based on charges of elitism with the added claim that society has a need for progress—are the widespread objections of developers and the masses of people seeking new dwellings in traditionally rural areas that sometimes approach the level of wilderness. For example, they seek to expand Californian housing developments steadily creeping into the desert or Brazilian slash-and-burn small, temporary farms steadily encroaching the Amazonian rain forest.

These differences of opinion among urban dwellers; between urban and

114

rural dwellers; and between advocates of wilderness and those seeking to develop new dwellings in largely undeveloped areas are not cool or detached at all. Indeed, they ensue in heated discussions and controversy, when not outright confrontation and mere conflicts of demands, and, together with the conflicts of which they are part and parcel, they raise a variety of questions. Salient among these are: Can a sound balance be found between the concerns these conflicting opinions and demands reflect? What natural or social facts set constraints on such balance?

These questions will provide the main focus for this essay. In addressing them, I will not seek answers to narrow policy questions—say, whether there should be new zoning regulations in Valdez, New Mexico, or whether wetlands legislation is soundly applied in the north end of Cheshire, Connecticut. Instead, I will seek to outline the interconnections of natural, economic, and cultural constraints within which policies and practices concerning urban and rural development need to proceed for given societies and the planet to remain viable.

I will put forth five theses. First, natural, economic, and cultural constraints affecting policy making, actions tending to alter societal practices, and even actions whereby individuals aim at making room for themselves and their aims in society, are characteristically shaped by the histories of development in the societies in which they are operant. Second, the development histories of these constraints lead to social ecologies, that is, to special adaptations and networks of interconnections between social groups, their habitats, and other groups, as well as to special adaptations and interconnections between these groups' individual members, their group or other groups, and their habitats. Third, given their significantly long-standing historical roots, these constraints can typically undergo only incremental changes that take long periods of time. Exceptional, more drastic and rapid changes are feasible when a variety of—often international—events somewhat external to the social ecology involved bring it close to its collapse. Fourth, despite their relative inflexibility, the constraints set by a given social ecology can make room for a great variety of policy-making, practice-changing, and personal development alternatives. Fifth, effective and morally defensible policies and decisions rely on and reinforce existing social ecologies instead of undermining them, except when these ecologies are likely to be found morally intolerable to any individuals affected by them who are of sound and cool mind, free from the influence of coercion and manipulation, and as informed as circumstances permit, or when there is clear and present danger that the ecologies will collapse, rendering the policies and decisions that rely on them inoperant.

In attempting to establish these theses, I will begin by examining the historical development of two social ecologies in some detail. One, the settlement of the pampas—the plains of Argentina, Uruguay, and southwestern Brazil—concerns the development of a social ecology in a largely unsettled region. The other, the creation of the interstate highway system in the United States, concerns the development of a social ecology within a significantly developed, indeed, industrialized society. Throughout this essay, in order to indicate the general nature of my conclusions, I will be indicating the similarities between the cases discussed and a variety of other cases.

UNSETTLING SETTLEMENTS

The Settling of the Pampas

An Early Ecology of International Trade

The shortest topographical distance between two points is rarely compatible with the shortest social distance between people's needs or wants and their satisfaction. It often fails to help and, indeed, hinders the satisfaction of these needs or wants. A case in point can be found in the pampas, the plains of the southern cone of South America. Early during the settlement of the pampas, traffic was of no serious concern. Indeed, at the time of the first founding of what today is the city of Bueños Aires, which is reported to have taken place in 1536 by Pedro de Mendoza, indigenous peoples were scarce nomadic hunters in the pampas.[2] The land had been a real wilderness—the remnants of old marshes, where *ombúes* had floated and now raised their gigantic bush branches in a land without trees, inhabited by many and varied species, but neither cattle, nor sheep, nor donkeys, nor horses, nor any other animals typically used as resources in Europe at the time.[3]

The fort Mendoza left in Bueños Aires upon returning to Spain was under "Indian attack and torn by internal dissent"[4] during the year it took him to return. Many died and its survivors eventually abandoned it in 1541, moving up the estuary and the Paraná River to an advance post, which today is Asunción, Paraguay. There, the numerous indigenous population was semisedentary, living in villages and growing crops, thus making it possible for Europeans to find a steady and reliable source of food. All in all, about 350 Europeans, including some Portuguese, Germans, and Italians, started the Spanish colony of Paraguay in Asunción.[5]

The developments just discussed already evidence a fact of social ecology. In order to establish the initial colonies, the Europeans arriving with the sixteenth-century Spanish expeditions to the southern cone of South America needed sedentary or semisedentary indigenous populations, at least because these populations offered a geographically fixed and recurrent source of food.

In addition, the Spanish expeditions included few artisans, a minimal number of blacks, and hardly any European women. This led to a society in which so-called Spaniards were often mestizos. It also led both the few, actual Europeans and the mestizos to adopt the food, customs, and even the language, Guarani, of the local indigenous population. On the other hand, Hispanic ideals and methods worked themselves thoroughly into the indigenous population through the same channels by which indigenous influence affected Europeans.[6]

Bueños Aires was reestablished by Juan de Garay, with the support of the Spaniards of Asunción, in 1580, and it began to thrive as a center for international commerce with Europe. To be sure, leather, a product that eventually became a symbol of the region's economy, was traded; but it was insignificant compared with the slaves, ironware, sugar, and luxury items that were exchanged for silver from (and most of which were transported to) Potosí, in modern-day Bolivia. A new social ecology based on international commerce was beginning to develop. In the process, the routes for moving merchandise across the plains began to be settled.[7]

These developments continued between 1580 and 1640, a period of Iberian union whose major agents in the South Atlantic were the Portuguese: "Bueños Aires was in effect a Portuguese 'factory' for trade with Peru."[8] In the meantime, the 1535 Bueños Aires's garrison horses and cattle that had taken to the plains had multiplied, prompting additional significant changes in the culture and social ecology of the region. The indigenous peoples who moved about the pampas learned to tame horses. A new group of people, the gauchos, began to develop. They were largely mestizos and somewhat nomadic, partly living off the cattle they followed across the land. By the end of the 1700s, the gauchos would form a distinct subculture of southern Brazil and the Argentine-Uruguayan pampas.[9] Indeed, the first known mention of the word *gaucho* dates from 1774.[10]

The significance of cattle for the economy of the area at the time should not be exaggerated. To be sure, the cattle stock around 1778 is estimated to have been 1,000,000 to 1,500,000 heads.[11] But in addition to cattle, sheep, donkeys, and, of course, horses were common. More important, besides large ranches, there were smaller tracts of land, which amounted to farms. Many of them were primarily devoted to growing wheat, corn, barley, and horticultural products.[12] But even all of these products taken together were not the primary sources of income for the region. Consider the area surrounding Bueños Aires. Between 1779 and 1784, leather and other agricultural products, though beginning to grow in significance as export goods, amounted to only 14.6 percent of the value of Bueños Aires exports through the Atlantic, while metals from the Peruvian and Bolivian mines amounted to more than 82 percent of the value of those exports. Indeed, recognizing Bueños Aires' growing influence, in 1776, the Spanish crown had created the viceroyalty of

the Rio de la Plata with Bueños Aires as the capital and the silver mines in
Potosí under its jurisdiction. This made Bueños Aires, rather than Lima, the
main outlet for Potosí's silver exports.[13]

A Growing Ecology of Production and International Trade

The emphasis on silver exports described in the previous section, however,
was about to change drastically in forty years, leading to a new and more
complex social ecology of the pampas. Having staged a revolution in 1810,
when the king of Spain was imprisoned by Napoleon, and having declared
independence in 1816 once Napoleon was defeated and the king of Spain,
free again, returned to power, the United Provinces of the Río de la Plata
were now free from colonial restrictions. However, they were cut off from
Peru and Bolivia by still-raging independence wars. Further, responding to
world demand and local conditions facilitating agricultural production, the
Bueños Aires area dramatically shifted its emphasis from international
exports of Peruvian and Bolivian metals toward international exports of local
cattle and agricultural products. Significant among them was leather, which,
with the growing development of the industrial revolution in Europe, was
needed not just for shoes and saddles but, increasingly, for the moving parts
of machines.[14] As a result, in the 1820s, cattle leather amounted to 65 percent of the total exports through the Bueños Aires harbor.

Roads and the Incipient Urbanization of a Wilderness

With the appearance of fences in the 1800s, and the expansion of cattle
industry and agriculture, the Argentine rural environment began to be
urbanized. Towns developed in places where forts or trenches had been built
to keep the aborigines at bay.[15] And the road system was expanded. To a significant extent, the Argentine road system was built for military reasons.
First, the main roads connecting Bueños Aires with the Argentine northwest
were originally built during the Spanish invasions that came from Potosí,
Bolivia, to Tucumán in northwestern Argentina, where, in 1553, the
Spaniards initially established a supply center of agricultural products of
European origin for Bolivia. This development later led to building the road
from Bueños Aires to Tucumán, mainly with the participation of the Por-
tuguese interested in trading with Potosí.[16]

Second, the main roads connecting Bueños Aires with the Argentine
northeast were largely built and consolidated during the wars of independence in the 1800s, when the forces led by General Belgrano attacked the
Spanish forces that were moving in from Asunción, Paraguay. Before that,
the main route used to connect Bueños Aires with Asunción and other cities

founded along the shore of the Paraná River, such as Santa Fe and Corrientes, had been the river itself.

Third, the main roads connecting Bueños Aires and western Argentina were developed during the later Spanish invasion attempts to connect contingents in the central Argentina city of Córdoba, founded in 1573 as a south-of-Tucumán supplier for Potosí, with those moving east from Chile into what today is Argentine territory. Only later, and mainly during the wars of independence, were roads built connecting Bueños Aires to Córdoba and, farther west, the Andean region.

Finally, the main roads connecting Bueños Aires with the Argentine south were originally built during the 1880s so-called Conquest of the Desert, which in effect was an invasion aimed at expelling the South American Indian groups that had moved into Patagonia fleeing the Spanish invasion of Chile and at making sure that Chile did not take over Patagonia.

The Consolidation of the Modern Social Ecology

The significant military component in building the Argentine main road system has led to consequences undermining Argentine society still today. Some were economic. For example, building the roads through deserts and not through fertile areas because, from a military standpoint, those were the shortest and fastest, hence preferable, routes, led to increased production costs. Such a situation was reinforced by British activities in the area after the independence wars. By and large, the British bought the land alongside the existing roads and built the railroads on it.[17] This created a transportation structure whereby all main roads led to Bueños Aires and, from there, no longer to the Spanish metropolis but to London or Southampton.

Other consequences were political. For example, the new structure reinforced the position of Bueños Aires as the new Argentine commercial metropolis that ruled over its internal colony—the rest of Argentina. This crucial economic position, like the position that led Chicago to become the prominent city it is in the United States, contributed to Bueños Aires's success. But, in contrast to Chicago, Bueños Aires was the only city of this nature in the entire country, and it had inherited the highly centralized bureaucratic-economic structure created by the Spanish crown, including its capital status. As a result, land came to be owned largely by people based in Bueños Aires, and the provinces became ever more dependent on the metropolis despite the constitution established in 1853, which was meant to create a federal state but, some argue, contributed to turn the provinces into the internal colony of Bueños Aires.[18]

In addition, the transportation structure, coupled with the developments just described, had ecological consequences. One was the almost complete elimination of forests in large areas of the Argentine northwest. They were cut down for Bueños Aires–centered development purposes.

The full set of components of the new social ecology being formed continued its settlement until well into the nineteenth century: A closely intermarried group of influential families, often descendants from colonial merchants, lived in Bueños Aires, controlled political office, and owned sophisticated large estates in the countryside. But a crucial element of the early social ecology remained: They never gave up involvement in the import-export business.[19]

Socioecological Constraints on Later Developments and Policy-Making Options

Argentine developments in the late nineteenth century and first part of the twentieth century cannot be understood except against the background of the previously described social ecology. For example, the masses of European—primarily Italian and Spanish—immigrants who came to Argentina between the 1880s and 1940s tended to stay in the Bueños Aires area. To be sure, other factors contributed to reinforce this phenomenon. For example, since the times of the Roman Empire, and throughout the Spanish conquest, it had been extremely difficult to keep people away from urban centers.[20] But crucial factors in Argentina were the unavailability of land and the concentration of economic opportunity in the metropolis.[21]

Many of Argentina's infrastructural difficulties through the nineteenth and twentieth centuries have been part and parcel of the complex social ecology whose initial development was previously described and whose interconnected components could not be soundly addressed in isolation from the others. That is, the imbalances could not be soundly addressed except within the overall constraints created by the said social ecology and the international conditions surrounding it. In this manner, the preceding discussion of the pampas's and, especially, of Argentina's development illuminates and provides some support for the first three of this essay's theses: First, natural, economic, and cultural constraints affecting policy making, actions tending to alter societal practices, and even actions whereby individuals aim at making room for themselves and their aims in society, are characteristically shaped by the histories of development in the societies in which they are operant. Second, the development histories of these constraints lead to social ecologies, that is, to special adaptations and interconnections between social groups, their habitats, and other groups, as well as to special adaptations and interconnections between these groups' individual members, their group or other groups, and their habitats. Third, given their significantly long-standing historical roots, these constraints can typically undergo only incremental changes that take long periods of time. Exceptional, more drastic and rapid, changes are feasible when a variety of—often international—events somewhat external to the social ecology bring it close to its collapse.

As indicated in the preceding discussion, the latter international events may be economic—say, a sharp increase in demand for cattle products resulting from the European industrial revolution; political—say, the imprisonment of Spain's king by Napoleon; or have some other nature. In any case, even drastic changes do not produce a complete discontinuity with the preceding social ecology. They only bring to prominence undercurrents, components of the social ecology—say, the cattle and agricultural industries in Argentina during the nineteenth century—which were in the background during previous stages of societal development.

One might believe that the case just discussed, at best, supports the essay's first three theses concerning the development of wilderness or undeveloped regions, but not concerning highly urbanized and industrialized areas. This belief, however, is false. In order to establish that it is false, let us turn to our next case.

The United States Interstate Highway System and the Automobile in the Twentieth Century

The Advent of the Automobile

When automobiles were introduced in the United States at the beginning of the twentieth century, nobody could have predicted the carnage they were going to bring upon the U.S. population. For this carnage was a significant result of the interstate highway system, whose construction was motivated by the cold war. And no data were then available to predict the cold war, hence the interstate highway system. Nor were any data available to predict the fact that the U.S. population was going to find the carnage tolerable enough not to outlaw the automobile or significantly restrict its use. This is not just a matter of unpredictability. A process involving various social decision procedures—most significantly, critical scrutiny and social interaction with the new technology and people affected by it—help work out morally significant details such as these.

The Social Assessment Process

No doubt, from the start, the process involves reasons for morally assessing policies and decisions. For example, upon cool and careful reflection, people may prefer policy arrangements that do not preempt future corrections. Say, concerning automobile use, people may prefer arrangements that would permit corrections when the seriousness of smog and highway fatalities becomes evident.

In addition, the parties involved may want to leave options open, so that these are worked out through negotiation, bargaining, and other social decision procedures. For example, they may want to engage in political dialogue, negotiation, and bargaining concerning the value customarily assigned to automobile use. This is not just a matter of gathering further evidence so that one can make more accurate predictions, or determine utilities more precisely, so that a sounder political judgment can be formed. That is, the utilities themselves are worked out in the process because people's attitudes and judgments about the alternatives are formed and transformed with the process. This process of critical scrutiny and social interaction leads to the settlement of reasons for assessing policies and decisions. It also leads to a cluster of natural, economic, and cultural conditions that are highly entrenched in people's lives and set constraints on what policies and decisions are viable for dealing, for example, with environmental problems related to automobile use.[22]

The Resulting Ecology of Automobile Use

Concerning automobile use in the United States, the social decision process has led to attitudes and judgments about such use which would have been unthinkable if all the information about the consequences of introducing automobiles in people's lives bad been known in advance. People have become significantly used to this independent, comfortable, and speedy means of transportation and have largely accepted the risks associated with it. Also, the automobile industry has become a crucial component of the U.S. economy. In addition, a variety of practices—such as carpooling—and policies—such as emission test policies—have become realities meant to address the said environmental problems. But even when the problems are huge, as with air pollution in Los Angeles, it has taken a long time and many failures of more conservative measures until policies were introduced to bring back a public transportation system that decades ago bad been dismantled to make room for freeways and motor vehicles.

Further, these measures have not been accompanied by restrictions on automobile use, because the social ecology connected with this use has crucial and hardly removable components—from individualistic attitudes about freedom, privacy, and driving one's own vehicle to economic interests concerned with motor vehicle sales and related jobs and profits—which doom more drastic measures to failure.

Socioecological Constraints on Current Developments and Policy-Making Options

This emphasis on individual freedom and privacy should not be taken lightly. As Mark Sagoff has convincingly argued in *The Economy of the Earth*, "An environmental policy . . . must attend to the meanings things have for us, for example, the meaning of pollution, not simply the magnitude of the risks it may pose."[23] Indeed, even if, in the case of motor vehicles in the United States, one may argue that environmental policy has emphasized freedom and privacy too much, one cannot soundly argue that it has not attended to the meaning of motor vehicles on the face of the pollution they cause and beyond the risks such pollution poses.

To do so is only realistic, hence, rational. For, whatever the right balance between freedom and privacy on the one hand and health risks from motor vehicle–caused pollution on the other, policy making about motor vehicles must proceed in accordance with the social ecology that involves these and many other moral and aesthetic concerns, as well as interests, of the population affected. That is, generalized practices and personal development aims are part and parcel of the social ecology. Hence, so long as the social ecology is viable, it is the path of prudence and good moral sense to change the said practices and pursue the said aims only within the overall constraints set by the prevalent social ecology.

Here again, in an industrialized setting, the natural, economic, and cultural constraints that centrally constitute social ecologies and affect policy making, practice change, and personal development actions are shaped by their histories and can undergo only incremental changes that take long periods of time. Hence, just as with the previously discussed development of the Argentine and neighboring plains, the preceding discussion of the U.S. interstate highway system illuminates and provides support for the first three of this essay's theses: First, natural, economic, and cultural constraints affecting policy-making actions tending to alter societal practices, and even actions whereby individuals aim at making room for themselves and their aims in society, are characteristically shaped by the histories of development in the societies in which they are operant. Second, the development histories of these constraints lead to social ecologies, that is, to special adaptations and interconnections between social groups, their habitats, and other groups, as well as to special adaptations and networks of interconnections between these groups' individual members, their group or other groups, and their habitats. Third, given their significantly long-standing historical roots, these constraints can typically undergo only incremental changes that take long periods of time. Exceptional, more drastic and rapid, changes are feasible when a variety of—often international—events somewhat external to the social ecology bring it close to its collapse.

As indicated in the preceding discussion, the said international events may be economic—say, a sharp increase in fossil fuel prices; political—say, the intransigence of the oil cartel; or have some other nature. In any case, they do not produce a complete discontinuity with the preceding social ecology but only bring to prominence components of the social ecology—say, the limited availability of electricity or other non-fossil-fuel-propelled vehicles in the United States—which were in the background during previous stages of societal development.

SOCIAL ECOLOGIES AND THEIR CONSTRAINTS ON POLICY MAKING, CHANGES IN PRACTICES, AND PERSONAL DEVELOPMENT

Dealing with Stable Situations

The cases just described are not isolated. Consider the United States's predominant agricultural practices, the widespread practice of relying on electric power in the United States and elsewhere, the practice of using airline transportation throughout the planet. They do not exist in a void but are deeply entrenched in social ecologies made up of natural and other circumstances, interests, wants, needs, ideas, and ideals. To be sure, many criticisms may be raised against the said practices. One can, for example, point out that deep tillage in U.S. agriculture leads to significant soil erosion; that the predominantly used sources of electricity—coal, oil, and nuclear—involve serious environmental and health risks; and that air travel is a significant source of pollution and resource depletion. But, in the main, for the foreseeable future, the U.S. agricultural system, as well as the widespread use of electric power and air travel on earth, are here to stay. For, as stated, they are deeply entrenched in social ecologies involving the economies, politics, and attitudes of widespread populations.

Despite the inflexibility of these constraints, however, a variety of changes can be brought about within their scope. Deep tillage is not always —if ever—unavoidable. Energy can be, and has been, conserved, and the sources of electricity can be, and have been, changed, for example, by using wind turbines or solar energy where the habitats and other circumstances permit. Attitudes can also change, as evidenced by the significant increase in preferences for carpooling in the United States and for public transportation in various densely populated areas throughout the planet.[24] This leads to our fourth thesis: Despite their relative inflexibility, the constraints set by a given social ecology can make room for a great variety of policy-making, practice-changing, and personal development alternatives.

More drastic changes may seem preferable to those dealing with the said environmental problems posed by radiation, pollution, and soil erosion. However, as I have discussed concerning policy making elsewhere, they are not preferable so long as the current social ecology makes them unfeasible or ineffective.[25] For, in such a case, they would be merely ideal and never come to be actual changes, and trying to bring them about would be a harmful distraction leading, at least, to a significant waste of time likely to turn previously feasible options for dealing with the problems into options that become unfeasible because, given the delay, time has run out on them.

Of course, for decision purposes, whether they are likely to be unfeasible or ineffective must be assessed in advance, not after the fact. For example, in retrospect, Greenpeace's attempts at stopping France's 1995 new nuclear tests turned out to be ineffective; but this is knowledge one has after the fact, hence irrelevant for assessing whether these attempts were preferable to other ways of expressing opposition to the French plans. Whether they were preferable, or even permissible, had to be established for decision purposes on the basis of grounds available before they were carried out.[26]

Dealing with Unstable Situations

A more drastic approach is in order when current social ecologies are about to collapse. For, in such a case, policies and decisions based on the current ecology are likely to be ineffective because their effectiveness is likely to go under together with the existing social ecology.

As for those situations in which, though still unshaken, a social ecology —say, the social ecology of slavery in various countries during the early 1800s—is morally *intolerable* to any of those affected by it, ways should be sought to change it.[27] First, if possible, these should be nonconfrontational, as those used in what today are Argentina and Uruguay, which involved hardly any opposition. Also nonconfrontational, and perhaps unavoidable, though, no doubt, not the most ideal from a moral standpoint, was the approach Brazil used to free the slaves in the nineteenth century. It was gradualist and always regarded as a response to what was considered a national, not regional, problem.[28] A second choice would be using confrontational but nonviolent methods as in Chico Mendes's approach to environmental problems in Brazil in the late twentieth century.[29]

Third, if none of the previous procedures is feasible or likely to be effective, then more confrontational procedures such as boycotts, strikes, and, in very extreme circumstances, even combat may be permissible, so long as they are accompanied by mediatory or other efforts aimed at stopping or diminishing the degree of confrontation and increasing the use of procedures that primarily involve appeals to reason and meaningful dialogue.

Of course, it is possible that such confrontational procedures will not be feasible or likely to be effective. In this fourth case, coping and waiting may still be used, as the main industrial countries of earth have done in the late twentieth century concerning China's and India's economic development based on industries that have chlorofluorocarbons as by-products and, hence, contribute to depleting the earth's ozone layer.[30]

Whatever procedures are appropriate for dealing with problems posed in given social ecologies, the preceding discussion provides reasons for this essay's last thesis: Effective and morally defensible policies and decisions rely on and reinforce existing social ecologies instead of undermining them, except when these ecologies are likely to be found morally intolerable to any individuals affected by them who are of sound and cool mind, free from the influence of coercion and manipulation and as informed as circumstances permit, or when there is clear and present danger that the ecologies will collapse, rendering the policies and decisions that rely on them inoperant.

SOCIAL ECOLOGIES AND PHILOSOPHY

The preceding discussion points to the crucial role of social ecologies in human life. Accordingly, I have been implying, neither an atomic conception of persons nor a corpuscular conception of social entities. Though not atomistic, the position in this essay does not ignore the crucial role of individual values and aims in social ecologies. Indeed, they often take precedence over other concerns. Besides, one social ecology can replace another and does not in anyway hold, as Whitehead once put it, that "every entity is only to be understood in terms of the way in which it is interwoven with the rest of the universe."[31] Hence, the position I have been espousing is not organicist.

I have simply argued that sound policy making, practice changing, and actions aimed at one's personal development involve seeking a *balance* between diverse and often conflicting individual aims and the social ecology in which the policies are made, practices changed, and personal development sought, provided that this ecology is stable. In this way, the preceding account is closer to that of David Hume than to that of Thomas Hobbes, at least in that, as Annette Baier puts it in discussing Hume's conception of the laws of nature, "conforming is preferred over non-conforming, so long as the sweets of coordination are achieved by conformity."[32]

In taking this stand, however, only the worst excesses of liberalism— those that lead it to oppose social integration—are ruled out, thus making it compatible with democracy. There are very good reasons for this. One reason is that reliance on reason and meaningful dialogue are seriously restricted, when not altogether excluded, under undemocratic approaches. In such circumstances, only arbitrariness and caprice remain, hardly the sort of thing likely to lead social groups or individual lives to survive, let alone flourish.

This leads to a second reason to curb the extreme, antisocial, and undemocratic forms of liberalism. As a result of characteristically bypassing reliance on reason and meaningful dialogue, such forms of liberalism breed conflict and confrontation *as a rule*. This is a recipe for disaster. For it leads to the dissolution of social groups into circumstances close, if not identical, to the state of nature Hobbes described as "solitary, poor, nasty, brutish, and short."[33] Hence, the said extreme, antisocial forms of liberalism are both politically unwise and morally objectionable.

As indicated at the end of part 2, it is fortunate when a social ecology that makes central room for moral practices and reasons is substantially operant. For it is the moral equivalent of the ozone layer: Without it, we are gone. It is, then, our ongoing task as policymakers, or simply as members of society, to help preserve and strengthen it and, when a given social ecology is likely to collapse, to help replace it with a new social ecology that also makes central room for reliance on reason and meaningful dialogue.

Whatever our response to Leonardo's dream city-block mentioned at the outset, we need to seek it through such social ecologies. For not just the humanities but, judging by what philosophers and historians of science seem to believe, also the sciences advance, not primarily because of their methods or principles, but primarily because they proceed in an openminded, flexible, critical, and self-critical manner.[34] Indeed, there is no better approach, because there is no other approach for us, who would cease to be ourselves in the only possible alternative: the utter forlornness of a suspended society and rampant state of nature.

NOTES

1. My source is a Spanish translation of some of Leonardo da Vinci's posthumous writings: Leonardo da Vinci, *Breviarios* (Bueños Aires: Editorial Schapire, 1952), Manuscrito B, 16r., p. 27.

2. James Lockhart and Stuart B. Schwartz, *Early Latin America: A History of Colonial Spanish America and Brazil* (Cambridge and New York: Cambridge University Press, 1983), p. 259.

3. Indeed, the term *pampa* meant "land without trees" in the language of the indigenous peoples of southern Chile, the *mapuches*, or people of the earth, whom the Spaniards called Araucans. Another etymological origin suggested for *pampa* is the Quechua term *bamba*, which means "plain" (see Jess Stein et al., *The Random House Dictionary of the English Language* [New York: Random House, 1970], p. 1041).

4. Lockhart and Schwartz, loc. cit.

5. Ibid., pp. 259–60; see also p. 88 for the dates of Bueños Aires's first founding and ulterior abandonment.

6. Ibid., pp. 259–62.

7. Ibid., pp. 272–73.

8. Ibid., p. 273.

9. The origin of the term *gaucho* is subject to much controversy. Some think it derives from the Arawak *cachu*, which means "comrade." See, for example, Jess Stein et al., op. cit., p. 587. Others suggest *gaucho* derives from the *guarani ca'úcho*, which means "tipsy." See, for example, José Gobello, *Diccionario Lunfardo* (Bueños Aires: A. Peña Lillo, 1978), p. 97. Still others have suggested that it derives from the southern South American Spanish term *guacho*, which means "illegitimate" or "orphan," or that it is related to the Provençal term *gauche*, meaning "twisted," "astray," "erring." See, for example, Magnus Mörner, *Race Mixture in the History of Latin America* (Boston: Little, Brown, 1967), p. 77 n. 9. Other scholars have pointed to the undoubted similarities between the bedouins and the gauchos—from their attire to their cattle-related activities—and accordingly hypothesized a north African influence in Spain that led to the development of Latin American gauchos. For this, see Mörner, ibid., p. 14. The linguistic fact that the term *gaucho* is similar to the Arabic *chaucho*, or herdsman, fits this latter hypothesis and has sometimes also been mentioned in etymological discussions.

10. See Mörner, op. cit., p. 77 n. 9.

11. Jorge Gelman, "Nuevas Imágenes de un Mundo Rural: La campaña rioplatense antes de 1810," *Ciencia Hoy* 1 (December 1989/January 1990), pp. 57–58. See also Lockhart and Schwartz, op. cit., p. 383.

12. Ibid., p. 58.

13. Ibid., p. 60.

14. Ibid. See also Lockhart and Schwartz, op. cit., pp. 336–41.

15. Ezequiel Martinez Estrada, *Radiografía de la Pampa* (Bueños Aires: Losada, 1953), p. 38.

16. Lockhart and Schwartz, op. cit., pp. 254–55.

17. This is evidenced by the frequent experience of those who travel across Argentina by train (or, alternatively, on the highways) and, almost invariably, see the highway running parallel to the railroad tracks (or, if traveling on the highway, the railroad tracks running parallel to it).

18. Martínez Estrada, op. cit., pp. 186–91.

19. Lockhart and Schwartz, op. cit., p. 341.

20. James Lockhart, *Spanish Peru, 1532–1560: A Colonial Society* (Madison and Milwaukee, Wis., and London: The University of Wisconsin Press, 1968), p. 21.

21. See, for example, Jacques Lambert, *Latin America: Social Structures and Political Institutions* (Berkeley and Los Angeles: University of California Press, 1969), p. 67ff.

22. This case has been discussed strictly at the policy- and decision-making level in my *Philosophy as Diplomacy*, pp. 32, 33, 161.

23. Mark Sagoff, *The Economy of the Earth* (Cambridge and New York: Cambridge University Press, 1988), p. 128.

24. Changes in preferences for public transportation are evident, for example, in Bueños Aires, where a growing number of people leave their motor vehicles home and use one of the subway lines, trains, and innumerable bus lines available to go downtown. A similar situation can be found in the New York City area. Even in the sharply different Los Angeles area, public transportation is being developed after having been relegated to the background in the past. An interesting comparison of Los Angeles and Bueños Aires can be found in David White, "When Is Los Angeles Going to Grow Up?" *The Dispatch* 23, no. 1 (January 1992): 1, 3.

25. For a detailed discussion of these points at the policy- and decision-making level, see my *Philosophy as Diplomacy*, pp. 140–41.

26. Ibid. Concerning the Greenpeace affair, see, for example, Patrick Dilger, "Yale Prof Involved in Suit Against France," *New Haven Register*, Thursday, October 19, 1995, p. A3.

27. *Philosophy as Diplomacy*, pp. 140–41.

28. Slavery was abolished in modern-day Argentina and Uruguay almost immediately after the May 25, 1810, revolution. Indeed, except for Brazil, Latin American countries abolished slavery fairly soon after independence. See, for example, Lambert, op. cit., p. 67. As stated in the essay's main text, Brazil's elimination of slavery was gradualist and a response to what was always regarded as a national, not regional, problem. For a brief but illuminating discussion of the history of slavery in Brazil, see Frederick B. Pike, *Latin American History: Selected Problems* (New York: Harcourt, Brace & World, 1969), pp. 134–70.

29. For details of this case, see Andrew Revkin, *The Burning Season* (Boston: Houghton Mifflin, 1990), and Alex Shournatoff, *The World Is Burning* (Boston: Little, Brown, 1990). For a philosophical discussion of this case, see Lisa H. Newton and Catherine K. Dillingham, *Watersheds* (Belmont, Calif.: Wadsworth, 1993), pp. 135–53. See also the discussion of this case in Chapter 4 of this book "Ecodiplomacy: Ethics, Policy Making, and Environmental Issues."

30. *Philosophy as Diplomacy*, p. 226, n. 29, 30.

31. Quoted in Paul Schilpp, ed. *The Philosophy of Alfred North Whitehead* (Evanston, Ill.: 1941), p. 687.

32. Annette C. Baier, *A Progress of Sentiments* (Cambridge, Mass., and London: Harvard University Press, 1991), p. 233.

33. Thomas Hobbes, "Leviathan," in D. D. Raphael, ed., *British Moralists 1650–1880* (Oxford: Clarendon Press, 1969) 1:37.

34. See, for example, John Dewey, *The Quest for Certainty* (New York: Putnam, 1960), passim. See also Thomas S. Kuhn, *The Structure of Scientific Revolutions* (Chicago: University of Chicago Press, 1962; 2d ed., 1969), passim; Imre Lakatos and Alan Musgrave, *Criticism and the Growth of Knowledge* (London: Cambridge University Press, 1974), passim.

Dialogue

— I am uncomfortable with the element of conservatism in your position.

— What element is that?

— Your siding with prevalent social ecologies except in very extreme circumstances.

— This is not my position.

— Please explain it then.

— I will gladly do so. First, my position rules out attempts at destabilizing a prevalent social ecology *head-on*, when the success of such attempts is quite unlikely and, given the available evidence, they will probably be ineffectual and even backfire. But ruling this out in no way amounts to siding with the prevalent social ecology.

— So, your account does not rule out working toward eroding prevalent social ecologies?

— My account only rules out recklessness. Indeed, it does not rule out revolutionary changes; but, for them to be justified, certain conditions must be met.

— What are they'?

— I have described them in essay 8 of *Philosophy as Diplomacy* concerning policy and decision making. But, through generalized decision making, they also apply in our wider context where conventions are settled. Here, I can give you only some brief examples. First, it is permissible and arguably prudent to go against a prevalent social ecology when there is clear and present danger that it will collapse. For, even if such ecology had morally appealing characteristics, it would be feckless to plan for a future predicated on its existence. One should, in such cases, seek viable and morally acceptable alternatives.

— Another case is that of morally intolerable social ecologies, say, those involving slavery. If likely to be both feasible and effective, it is permissible, if not required, to go against such social ecologies, even if this involves sacrifices on the part of their victims so long as, when of sound mind, free from the influence of coercion and manipulation, and well informed, these victims or their representatives would find the sacrifices worthwhile. In other words, when likely to be feasible and effective, undermining highly oppressive social ecologies is permissible, so long as it does not amount to attempting to free the oppression's victims against their own competent and informed will.

— What if they object?

— Then I should act without arrogance. Under the conditions described, those people should be the best judges of what should be done about their situation.

— Would you then side with the prevalent social ecology?

— I need not do that. Indeed, I might still oppose it, and it would be quite permissible for me to engage in a variety of social decision procedures, from discussion of merits to consensus building in order to change things for the better. But it would be impermissible to act so as to force people to be free or sacrifice individuals to a cause, however appealing and ideally just the latter might appear to us. If this is conservatism, what is wrong with it?

— Even if your position is sound concerning economic and political conditions, I cannot see how it could be extended to other areas of human activity. How would it apply in art, science, or philosophy? By the way, you have mentioned social ecologies many times but said little, if anything, about philosophical ecologies and their relation to other ecologies. This is odd, given that the book's title is *Philosophical Ecologies*.

— You are right. I should give a detailed account of philosophical ecologies, and how they relate to other intellectual—say, scientific—ecologies, as well as to social ecologies. I should also explain what place they have and what role they play in the activity of philosophy, in other words, in philosophical inquiry. I will do this, beginning with art and literature, in the coming essays. Please be patient.

— I will be patient, but not easily convinced.

— That is how things should be. Let us proceed.

Fantasizing Realities:
Art, Literature, and
Cross-Cultural Aesthetics

INTERPRETING TEXTS

The Argentine writer Jorge Luis Borges once said: "Each reading of a book, each rereading, each memory of that rereading, reinvents the text."[1] One might take this to be the preposterous claim that the book's text understood as the list of its sentences—that is, the list of grammatically complete series of linguistic symbols constituting the book—changes with each reading, rereading, or recalling of these acts. No doubt, there is some reason for attributing this interpretation to Borges's statement. After all, Borges is known to have had a penchant for defending preposterous claims.[2] But, in this case, he was making a different—some would argue, an equally preposterous—claim that went beyond syntax into a book's meaning and function:

> When the book lies unopened, it is literally, geometrically, a volume, a thing among things. When we open it, when the book surrenders itself to a reader, the aesthetic event occurs. And even for the same reader the same book changes, for we change.[3]

No doubt, we do change. Yet why should the book and its text change with each reading? Why not say that neither the book nor its text, but only our interpretation of these, changes with each reading? And if the book and the text change as suggested, how could it be possible to appraise soundly, hence pass value judgments on, communicate about, and even remember them? Indeed, by analogy with books, how could it be possible to appraise soundly, communicate about, and even remember any product of human creativity? These are the questions I will address in this essay, with special regard for cross-cultural interpretations of works of art and literature.

I will pursue seven theses. First, whatever the aesthetic value of given items—from sculptures and paintings, through poems and plays, to buildings and natural landscapes—it is not always a matter of aesthetic experience or, for that matter, of cognition; but it is functional in a general manner that makes the provision of aesthetic experiences crucial in some cultures and the promotion of such things as social cohesiveness and safety crucial in others, Second, the intention of the author of a work of art or literature is not decisive in evaluating this work. Third, the intention of the author of a work of art or literature is not decisive for its being art or literature or for its having aesthetic value. Fourth, whether given items are works of art or literature is settled through an open-ended social process of critical scrutiny of the items (say, the works and their purposes), as well as interactions between these items and those who interpret them (their authors, if any, included), who sometimes engage in dialogue with each other in the process. Fifth, the values of works of art or literature, of many other culturally valuable products, and, indeed, of aesthetically valuable natural objects, are the result of open-ended dialogues. Sixth, despite the open-ended nature of the dialogues that lead to the identification and evaluation of works of art or literature, culturally valuable products, and aesthetically valuable natural objects, aesthetic appreciation and judgment are objective in that they are open to critical scrutiny by appeal to reason. Seventh, social ecologies that significantly involve interested, thinking, and appreciating individuals are crucial to the existence of art, literature, culture, and, indeed, valuable natural objects.

A MATTER OF DIALOGUE: ARTWORKS, LITERARY WORKS, AND AESTHETIC OBJECTS

During recent decades, the possibility of appraising artworks (and, indeed, literary works) has been thought increasingly problematic, partly because the notion that they have aesthetic value has been thought increasingly problematic. In "In Defense of Aesthetic Value," Monroe C. Beardsley puts the matter as follows:

> Although the philosophy of art flourishes in our day as never before, and although the word "aesthetics" is widely accepted as a label for this subject, the concept of the aesthetic has grown more problematic with the progress of aesthetics.[4]

Beardsley is particularly concerned with challenges to the possibility of soundly appraising—hence passing value judgments on—artworks. In response to these challenges, he defends the functional account of art appraisal, "which assigns to the art appraiser the task of estimating how well the artworks that fall under his scrutiny fulfill their primary or central purpose."[5]

This poses various questions: What criteria serve to establish what is the primary or central purpose of a given artwork? Take any given creation, what criteria serve to establish that it is a work of art? And what criteria serve to establish to what kind of art it belongs? Beardsley responds by characterizing an *artkind-instance* as "anything that belongs to some recognized artkind,"[6] and an *artwork* as "an intentional arrangement of conditions for affording experiences with marked aesthetic character."[7] He also comments:

> The terms "artwork" and "artkind-instance" are not coextensive, since some artkind-instances were not produced with the requisite intention, and some things produced with the requisite intention belong to no established artkind.[8]

One may wonder why Beardsley would want to draw such a distinction. He says that this distinction allows us to characterize art as having the purpose of providing experiences with a marked aesthetic character: "Art as a social enterprise is understood in terms of an aesthetic purpose (while allowing that very many artworks have other purposes as well)."[9]

Who is supposed to have this aesthetic purpose: the producer of the artwork, those who experience it, those who judge it, all of these, or some other individuals? Beardsley says,

> If and only if the intention to produce an occasion for experiences of this sort plays a notable part in the production of some object, event, or state of affairs will that thing be a work of art.[10]

But why should this be so? Take *Gulliver's Travels*. Its creator meant it to be a political satire, hence, to afford experiences of a specific aesthetic and cognitive sort. Historically, it turned out to be read in such a way that it is considered, in effect, turning out to be a children's book, affording aesthetic experiences of quite another sort. These latter experiences were never intended by the author, Jonathan Swift. Hence, though his intention to produce a political satire led him to write *Gulliver's Travels*, this intention played no notable part in the creation of the actual book, which came to be understood and evaluated in a way radically different from that he intended. Does that mean that *Gulliver's Travels* is not a work of art?[11]

One might reply that the preceding remarks too narrowly interpret the sort of experiences that must be intended for a particular creation to be a work of art. The author's intention must be to afford experiences with marked aesthetic value, even if they turn out to be of a specific sort unintended by the author. However, two questions arise: First, if it does not matter what specific aesthetic experiences the author intended to provide, then why should it be necessary that the author intended to provide any aesthetic experiences? Second, even if the general intention to provide some aesthetic experience were involved in leading an author to create a certain item, would this suffice for the creation's being a work of art or literature? I will address these and related questions in the essay's remainder.

AESTHETIC OBJECTS AND CULTURAL DIFFERENCES

Five main criticisms have been advanced against the position exemplified by Beardsley: functionalism in aesthetics. First, we ought not to judge artworks at all. Second, if we may judge, there is little or no point in doing so. Third, even if there is a point in judging, our judgments cannot be true or false. Fourth, even if they can be true or false, no reasons can support them. Fifth, even if reasons can support them, the judgments can never amount to knowledge.[12]

Some clues useful for addressing these criticisms, as well as for establishing whether providing experiences is somehow central to artworks, have been given by the contemporary African philosopher Innocent C. Onyewuenyi. Onyewuenyi criticizes the traditional, highly individualistic, and static Western approaches to aesthetics. By contrast, he says, "Existence-in-relation, communalism, being-for-self-and-others sum up the African conception of life and reality." And, given this culturally entrenched conception, "African art is functional, community-oriented, depersonalized, contextualized, and embedded."[13]

As for what the terms *functional* and *community-oriented* mean, Onyewuenyi says:

> By functional and community-oriented we mean that African arts—visual, musical, kinetic or poetic—are designed to serve practical, meaningful purpose, beauty of appearance being secondary. Now, if this account is accurate, the Western distinction between arts and crafts may well not apply to African arts. A carving, for example, is aesthetically beautiful in the African standard if it functions well as stimulus in the worship of the deity, the community of worshippers being the judges. Furthermore, the artist's individual purpose "is not to depict his own individual whims and feelings. . . . He performs rather in such a way as to fulfill the ritual and social purposes of his community."[14]

This points to our previous question: Why should it be necessary that the author intended to provide any aesthetic experiences? In other words, is the capacity to provide experiences somehow central to all artworks? The highly practical, not experience-oriented, nature of African artworks and art undermines Beardsley's notion that the function of art is to provide *experiences* with a certain distinctive, aesthetic character, which is central to their aesthetic value. This is not the function of African art, hence, not the function of art in general, if there is one.

On the other hand, by closely connecting the social functions of artworks with a variety of societal needs, from those that are ritual to those that are highly practical, African art comes to Beardsley's rescue, insofar as he is a functionalist, by undermining the five criticisms of functionalism previously mentioned. These criticisms might make some sense when the aesthetic experiences in artworks are sharply separated from the practical functions of art objects. However, when, as in African art, the separation collapses, the views become questionable. Let us see why.

First, in the social circumstances and roles within which artists work in Africa, one may, and ought to, judge whether an artist's product fits community needs or not. Hence, it is false that one ought not to judge artworks. Second, there is very much of a communal point to judging. Therefore, it is false that there is little or no point to judging. Third, the judgments can be true or false. For example, they are true when they state that the artwork meets community needs when it indeed does. So, the third criticism is false. Fourth, reasons of a significantly factual nature can often support these judgments. Hence, not only the judgments can be true or false, but good reasons can be given for or against them. Finally, when the judgments are true and the judgers are sure of what they say, and their reasons are good, then the judges have knowledge. Hence, it is false that the judgments cannot amount to knowledge.[15]

Going back to the relation between providing experiences and the purpose of art, to say that the capacity to provide such experiences is not the characteristic purpose of art (because not the purpose of *all* artworks) is not to say that no artworks can have this purpose. They can, and some Western artworks have it; but it is a nonessential, however culturally entrenched, purpose. It follows that Beardsley's criticism of Nelson Goodman's position does not have the implications Beardsley attributes to them. Beardsley says:

> Nelson Goodman's position rejects aesthetic value, as I have been analyzing and defending it, and proposes instead to base art appraisals on the cognitive value of artworks—their capacity to contribute to the "creation and comprehension of our worlds." Artworks turn up in this account as characters, or classes of characters, in symbol systems, and like other symbols are to be judged primarily, or centrally by their successful functioning as symbols, their "cognitive efficacy."[16]

Beardsley formulates various criticisms against this thesis. First, he argues that instrumental musical works and nonrepresentational paintings have not been shown to be such symbols, because it has not been shown that they refer to anything. Second, many natural objects, such as mountains and trees, seem to have a value closely akin to that of artworks; yet they are not characters in any symbol system. Third, cognitive concerns are often sacrificed in the arts for the sake of something else.

These seem to be good reasons against Goodman's position, but Beardsley thinks they also give credence to the notion that "the primary purpose is the aesthetizing of experience."[17] They do not. Take African masks used in dance. In this regard, Onyewuenyi quotes Peggy Harper:

> Through his dance Efe has the power to please the witches and so turn their malevolent self-seeking power into a generous benevolence towards the community.[18]

As previously argued, and further evidenced by Harper's statement, the aesthetizing of experience is not the primary purpose in African artworks. Instead, the provision of tools for social purposes, such as cohesiveness or safety, that at least sometimes are primarily noncognitive and not a matter of experience is its main function. As for Goodman's claims, no doubt the masks have to depict divine powers correctly; but, as stated, their overriding purpose goes beyond such cognitive aim. In any case, if, as Beardsley claims, cognitive claims are often sacrificed in the arts for the sake of something else, it need not be the case that this something else is an experience. It may be, and sometimes in fact is, a social tool intended to provide such things as cohesiveness and safety.

Is this position undermined by Beardsley's claim that many natural objects have a value closely akin to that of artworks? It would be if the social function just mentioned were claimed to be the primary purpose of all artworks. But I am advancing no such view. Instead, the position I am taking is that the particular nature of what has aesthetic value—whether a work of art or literature or a natural object—varies across cultures and, indeed, may vary within the Western tradition more than Beardsley seems to believe. Hence, there are good reasons for inferring this essay's first thesis: Whatever the aesthetic value of given items—from sculptures and paintings, through poems and plays, to buildings and natural landscapes—it is not always a matter of aesthetic experience or, for that matter, of cognition; but it is functional in a general manner that makes the provision of aesthetic experiences crucial in some cultures and the promotion of such things as social cohesiveness and safety crucial in others.

One might accept this thesis and still argue that the author's intention—whether it aims at providing aesthetic experiences, social cohesiveness, or something else—is crucial for the author's creation to be, and to be evaluated as, an artwork. I will next turn to a discussion of this view.

What Is Aesthetically Valuable?
A Matter of Dialogue

Authors' Intentions and Aesthetic Evaluation

How significant, if at all, for evaluating an artwork is what an author intended by the artwork? William K. Wimsatt and Monroe C. Beardsley raise this question in "The Intentional Fallacy." They argue that art criticism is not author psychology and has rules of evidence different from those of author psychology. Concerning poems, for example, they write:

> The paradox is only verbal and superficial that what is (1) internal is also public: it is discovered through the semantics and syntax of a poem, through our habitual knowledge of the language, through grammars, dictionaries, and all the literature which is the source of dictionaries, in general through all that makes a language and culture; while what is (2) external is private or idiosyncratic; not a part of the work as a linguistic fact: it consists in revelations (in journals, for example, or letters, or reported conversations) about how and why the poet wrote the poem—to what lady, while sitting on what lawn, or at the death of what friend or brother.[19]

The authors grant that there is "an intermediate kind of evidence about the character of the author or about private or semiprivate meanings attached to words or topics by an author or by a coterie of which he is a member,"[20] which is relevant to understanding the particular semantics of a poem. Yet, to use such evidence need not amount to engaging in author psychology:

> The use of biographical evidence need not involve intentionalism, because while it may be evidence of what the author intended, it may also be evidence of the meaning of his words and the dramatic character of his utterance.[21]

In support of this view, the authors reason as follows:

> The meaning of words is the history of words, and the biography of an author, his use of a word, and the associations that the words had for him, are part of the word's history and meaning.[22]

A cluster of questions arises here: Are the associations a person has for any word, however peculiar they may be, part of the history and the meaning of the word? Should the person be a competent user of the language to which the word belongs? Should the person also be an author or even a recognized author? In any case, in what manner is evidence of this kind to be used together with our habitual knowledge of the language in the criticism of the author's poem?

Though the authors' example is that of a poem, their position applies generally to both works of art and works of literature. How does it apply to the criticism of a painting? Take Pablo Picasso's work. Between 1901 and 1905, during the so-called blue period, the tonality of his paintings was predominantly blue but, other than that, he was still working within the constraints of pictorial realism. By 1910, however, his paintings began to involve basic, simplified forms and a markedly shifting point of view, which soon came to be called *cubism*. Suppose it is 1911 and one tries to engage in criticism of Picasso's 1911 painting entitled *Accordionist*. How could one fruitfully appeal to the existing habitual knowledge of painting, or even of Picasso's previous repertoire, to engage in such criticism? One might not even recognize the painting as a Picasso!

Is this, however, a good reason for concluding that the author's intentions are crucial for assessing the author's work? Not by itself. Let us consider *Gulliver's Travels* again. It was intended to be a political satire, but it turned out to be a children's book and to be evaluated as such. In this regard, its author's intention that the book be a political satire is hardly relevant. Hence, we have reason for stating our second thesis: The intention of the author of a work of art or literature is not decisive in evaluating this work.

What is decisive then? An alternative is this: Until the interpreters presented with a new kind of creation an artist has produced have predominantly settled on the matter, the judgment is still out on whether it is good art. Indeed, the judgment is still out on whether it is art at all. Along these lines, one could argue that, in the case of Picasso's cubist works, through a process of dialogue among artists and nonartists and through their interaction with cubist objects, the matter was settled, not by Picasso's intentions, but by those affected by Picasso's work—Picasso included—later in the century. This also applies to the previously discussed example of *Gulliver's Travels*. Though meant as a political satire, given the process of dialogue and interaction that followed its publication, it turned out to be a children's book with a certain literary value, however arguable this value may remain.

The Sociological Fallacy Objection

One might object that the position just formulated avoids the intentional fallacy by falling prey to a sociological fallacy. For it appears that, in this view, not the author's intentions (as in the intentional fallacy) but society's interpretations (hence the sociological fallacy) determine what is art, what is good art, and what kind of art an artist's particular creation is. Yet, this is false, because what society thinks at one time may, and often does, turn out to be different from, when not the opposite of, what society thinks at some other time.

This objection, however, is exaggerated. For, first, to say that the process of critical scrutiny of, and interaction with, the products of artistic activities is central to settle these matters is not to say that society does it all. Artists and their creations are an integral part of the criticism process. Indeed, without the artists' products, the process could not take place or take the direction it takes.

Jorge Luis Borges, the Argentinian author, has provided an example. In "Kafka and His Precursors," he considers three mutually dissimilar writings—Zeno's paradox against movement, a paragraph by Han Yu, a prose writer of the ninth century, and the works of Kierkegaard—and examines their relation to Kafka's work.

> If I am not mistaken, the heterogeneous pieces I have enumerated resemble Kafka; if I am not mistaken, not all of them resemble each other. This second fact is the more significant. In each of these texts we find Kafka's idiosyncrasy to a greater or lesser degree, but if Kafka had never written a line, we would not perceive this quality; in other words, it would not exist.[23]

That is, the artist's creation provides the referent for, and guides, the process of criticism. What it does not do is to settle matters in and of itself regarding whether the item being considered is art (or literature) or whether it is good art (or good literature). Hence, we can infer our next three theses: Third, the intention of the author of a work of art or literature is not decisive for its being art or literature or for its having aesthetic value. Fourth, whether given items are works of art or literature is settled through an open-ended social process of critical scrutiny of the items (say, the works and their purposes), as well as interactions between these items and those who interpret them (their authors, if any, included), who sometimes engage in dialogue with each other in the process. Fifth, the values of works of art or literature, of many other culturally valuable products, and, indeed, of aesthetically valuable natural objects, are the result of open-ended dialogues.

To be sure, the fact that all these items result from open-ended dialogues does not ensure that the dialogues will always converge on a shared or permanent view about the nature, kind, identity, and value of an artwork. But they often do. And the limited number of instances in which the dialogue may fail to attain closure provides no good reason to reject the view that artworks result from dialogues. This failure to attain closure can be a feature of some dialogues, either because the objects prompting them—for example, sculptures made by long-gone pirates on the coastal mountains of Ecuador—offer insufficient guidance to their interpreters, because interpreters cannot make up their minds about the objects—for example, about the large-scale drawings of bird-like and other stylized figures on the Peruvian Andes—or because later generations lose interest in the objects—for example, filetto drawings on Argentine trucks.

This latter possibility has the virtue of pointing out that art and literature, like culture and any cultural products, have to be regained by each generation. Classics can cease to be classics, and art forms, say, poetry, could cease to be art forms, just as a joke can cease to be a joke given sufficient cultural changes. By themselves, the objects that become classics, say, Homer's works, and those that are poetry, say, *The Odyssey*, are only the referents of the dialogues that turn them into classics and poems. If the dialogues ever cease, they can start again and bring classics and art forms back into existence. Indeed, as the institution of education makes plain, not just the classics but our entire cultural heritage, has to be recovered and redefined by each succeeding generation.

The Subjectivity Objection

At this point, a different objection could be advanced: The position taken in this section entails that artistic appreciation and judgment are an entirely subjective matter—a mere matter of taste. This objection, however, is mistaken. For, as I shall further explain later, at any point in the dialogue, it is possible for anyone to give reasons for or against the view that a given object is art, or art of a certain kind, or that it has this or that identity or value.[24]

The Fragmentation Objection

One might still argue that since, as made plain by our previous discussions and selection, art is highly fragmented in the twentieth century, such reasons are not always, if ever, possible. In other words, the said fragmentation is irreducible. This is a position relevant to what the Spanish philosopher José Ortega y Gasset discusses in *The Dehumanization of Art*, where he examines art and changes in artistic style from a sociological point of view.

Ortega acknowledges at the outset that "every newcomer among styles passes through a stage of quarantine. The battle of Hernani comes to mind."[25] However, he states that the unpopularity of present-day art is of a different kind. It does not result from a difference of tastes. "It is not that the majority does not like the art of the young and the minority likes it, but that the majority, the masses, do not understand it."[26] As a result, most people do not merely dislike the new art while, at the same time, feeling superior to it. They feel vaguely humiliated. Through its mere presence, the art of the young compels the average citizen to realize that he or she is just this—the average citizen. This arouses indignation. Yet, Ortega argues, "A time must come in which society, from politics to art, reorganizes itself into two orders or ranks: the illustrious and the vulgar."[27] He thinks the new art (Stravinsky's music and Pirandello's theater are his examples) is accessible only to a special class of human beings "who may not be better but who evidently are different."[28]

In what manner are these people different? Ortega argues that, in experiencing artworks, ordinary people have the same focus as in ordinary life: people and passions. "By art they understand a means through which they are brought in contact with interesting human affairs. Artistic forms proper—figments, fantasy—are tolerated only if they do not interfere with the perception of human forms and fate."[29] By contrast, Ortega argues, those who understand the new art are artists. Indeed, even if pure art is impossible, the new art "is an art for artists." By this, Ortega means "not only those who produce this art but also those who are capable of perceiving purely artistic values." This division, he adds, is insurmountable:

> Modern art . . . will always have the masses against it. It is essentially unpopular. Any of its works automatically produces a curious effect on the general public. It divides the public into two groups: one very small, formed by those who are favorably inclined towards it; another very large—the hostile majority.[30]

According to Ortega, this is so because, in contrast to ordinary people's preferences, modern art artists divest their objects of their aspect of lived reality. This is not to say that they do or should produce objects that have nothing to do with the ordinary objects of human life. They simply produce objects that, though resembling these ordinary objects, resemble them "as little as possible." As a result, such "ultra-objects" come across to ordinary people as dehumanized—mere deformations of ordinary life. But, Ortega argues, this is unavoidable:

> Perception of "lived" reality and perception of artistic form . . . are essentially incompatible because they call for a different adjustment of our perceptive apparatus.[31]

They are as incompatible as seeing a garden through a window and focusing on the windowpane at the exact same time. We cannot do both things, and, according to Ortega, perceiving modern art is like perceiving the windowpane, not the garden one can see through it.

One might respond that, though one cannot do both things at once, many people, certainly artists, can do one or the other at will. However, Ortega indicates that these are *socially* incompatible activities:

> Not many people are capable of adjusting their perceptive apparatus to the pane and the transparence that is the work of art. Instead they look right through it and revel in the human reality with which the work deals. When

they are invited to let go of this prey and to direct their attention to the work of art itself they will say that they cannot see such a thing, which they indeed cannot, because it is all artistic transparence and without substance.[32]

OBJECTIVITY IN DIVERSITY

No doubt, Ortega provides an accurate description of a type of fragmentation in the appreciation of art, literature, and, in general, items supposed to have aesthetic value or be culturally significant in the Western world during the first part of the twentieth century. Ortega's description does not entail, however, that such a fragmentation is forever unavoidable or irreducible. For, first, such irreducibility is just a hypothesis. Pointing to recurrent sharp differences in artistic appreciation between artists and members of the general public is not good ground for believing such differences are irreducible. As previously discussed, they do not appear irreducible, or even existent, concerning African art. Nor is it clear that they predominated, if they existed at all, during the Italian Renaissance. It may be true that now, given current Western linguistic, conceptual, stylistic, and practical differences among groups, the fragmentation is irreducible. But neither languages, nor conceptual frameworks, nor styles, nor practices are static. They can, and do, change through dialogue and social interaction, so that today's incommensurables are superseded by tomorrow's common ground.

Second, even if the construction of a common around of aesthetic appreciation were, at least on occasion, unworkable, the mutual irreducibility of points of view does not entail that artistic appreciation and judgment are an entirely subjective matter—a mere matter of taste. They would, if artistic appreciation and judgment entailed shared experiences. But they, at most, entail that good reasons could be formulated in the process of critical scrutiny and trial by practice. And such reasons, that supersede social differences, can be, and often are, formulated in the process. Hence, our sixth thesis: Despite the open-ended nature of the dialogues that lead to the identification and evaluation of works of art or literature, culturally valuable products, and aesthetically valuable natural objects, aesthetic appreciation and judgment are objective in that they are open to critical scrutiny by appeal to reason.

No doubt, without sufficiently interested, thinking, and appreciating individuals, the dialogues will tend to be unstable or hit dead ends. But this is simply to say that, without such individuals, the existence of art and artworks will be unstable and precarious, just as in a world with soft thinking and shallow individuals, the existence of knowledge and science, indeed, the existence of philosophy, will be unstable and precarious.[33]

AESTHETICS, SOCIAL ECOLOGIES, PHILOSOPHICAL ECOLOGIES, AND PHILOSOPHY

The previously described dialogue and social process of give-and-take that helps establish what is art, literature, culturally significant, and aesthetically valuable does not take place in a vacuum. Like the various processes discussed so far in this book, it is embedded in and partly prompted by the internal tensions of predominant social ecologies. These include interconnected interests, wants, preferences, and needs, as well as ideas about art, literature, and what is culturally or otherwise significant. These latter ideas are often also interconnected, as well as connected with other ideas about a variety of things, from everyday life to the entire universe. In this respect, they constitute outlooks on things, or philosophies, or, as I will say to emphasize their often interconnected and interactive nature, *philosophical ecologies*. In short, philosophical ecologies are networks of interactive ideas embedded in social ecologies.

It is in the midst of all these things that the process of dialogue and interactions described in this essay proceeds. In it, parallels can be found with the processes described in previous parts of the book. To begin with, here again, social ecologies play a crucial role. For example, they provide the standing conditions for meaningfully carrying out the social process of critical scrutiny and interactions—that is, for interpreting, communicating and criticizing, hence remembering oneself, one's interpretations and those of others, and the criticisms formulated along the way. As previously discussed, this process leads to settling what counts as art, literature, culturally or otherwise significant, and aesthetically valuable items.

Further, in this process and just as in other cases discussed in previous essays, there is a sense of philosophy that starts in the social and philosophical ecologies previously discussed but, through them, moves beyond them. This is philosophy as a critical, self-critical, and inquiring *activity*, which can make a contribution along the lines of pragmatic approaches, and the conception of philosophy as diplomacy, which I have developed elsewhere and already discussed in this book.[34]

In this regard, just as in previous essays, the position in this essay does not ignore the crucial role of individual values and aims in the process. Indeed, creative individuals are driving forces without which the process would become stale and, in effect, come to a dead end. Whatever our position on a specific art-related, culture-related, aesthetic, or philosophical problem or issue, we need to seek it through the socially embedded process of meaningful dialogue and interaction previously described. For not just the humanities or, as I will further argue in essay 9, also the sciences, but the arts and literature advance primarily by proceeding in an open-minded, flexible, critical, and self-critical manner.[35] Indeed, there is no better approach. For, without such social context for critical scrutiny and the inquisitive individuals motivated to engage in such scrutiny, there is no other approach that makes sense of the meaningfulness of art, literature, culture, and landscapes—a meaningfulness made plain by the care and concerns, when not controversy and confrontation, they elicit. This leads to our seventh and final thesis: Social ecologies that significantly involve interested, thinking, and appreciating individuals are crucial to the existence of art, literature, culture, and, indeed, valuable natural objects.

BETWEEN INDIVIDUAL DESOLATION AND SOCIAL STAGNATION

We started with Borges's statement that "each reading of a book, each re-reading, each memory of that re-reading reinvents the text." Given this essay's discussion, Borges's statement brings out a fact crucial to the process whereby works become art, literature, valuable as art of literature, culturally significant, or aesthetically valuable natural objects: the unavoidable ongoing recovery and redefinition of all these objects and values. It is also crucial to the process whereby some works of art or literature become and remain classics in the sense that they keep on passing the test of critical scrutiny and captivating people's imagination from generation to generation.

The said fact of ongoing cultural recovery and redefinition, however, is impossible in a social vacuum. For it to take place, the social ecologies that make critical scrutiny possible and meaningful are equally crucial. Without both, the individuals' critical scrutiny and the social ecologies within which it can only take place, not just art but also literature, philosophy, culture and its products in general, and our world of nature would be dehumanized beyond Ortega's conception of this phenomenon. We humans and our entire world would fall prey to a much more thorough dehumanization in the desolation of individuals cut off from all else or in the stagnation of social ecologies devoid of diversity, its concomitant tensions, and its frequent result: creative thought.

NOTES

1. Jorge Luis Borges, *Seven Nights* (New York: New Directions, 1984), pp. 76–77.
2. Alastair Reid, introduction to *Seven Nights*, by Jorge Luis Borges, p. 3.
3. Borges, op. cit., p. 76.
4. Monroe C. Beardsley, "In Defense of Aesthetic Value," *Proceedings and Addresses of the American Philosophical Association* (Newark, Del.: APA, 1979), p. 723.
5. Ibid., p. 724.

6. Ibid., p. 725.

7. Ibid., p. 729.

8. Ibid.

9. Ibid.

10. Ibid.

11. Jorge Luis Borges brings up this example, which had been Rudyard Kipling's, in discussing what he describes as "the error of supposing that intentions and plans matter a great deal." See "The Argentine Writer and Tradition," in Jorge Luis Borges, *Labyrinths* (New York: New Directions, 1962), p. 185.

12. Beardsley. op. cit., p. 731.

13. Innocent C. Onyewuenyi, "Traditional African Aesthetics: A Philosophical Perspective," *International Philosophical Quarterly* 24, no. 3 (September 1984): 242.

14. Ibid., pp. 242–43.

15. Here, I am assuming only the largely uncontroversial characteristics of what it is to know rather than merely to believe that something is the case. An individual knows something in this sense when the following three conditions are satisfied: (1) What the individual believes is true; (2) the individual is certain of what he or she believes; and (3) the individual has, in some sense, good reasons for believing it. Conversely, if these three conditions are satisfied, then the individual knows what he or she believes. As stated, all this is largely uncontroversial. The controversy arises only concerning what reasons, if any, are good reasons for being certain of what one believes.

16. Beardsley, op. cit., pp. 744–45.

17. Ibid., p. 745.

18. Quoted in Onyewuenyi, op. cit., p. 243.

19. W. K. Wimsatt Jr., and Monroe C. Beardsley, "The Intentional Fallacy." in W. K. Wimsatt, *The Verbal Icon* (Lexington: University of Kentucky Press, 1967), p. 10.

20. Ibid.

21. Ibid., p. 11.

22. Ibid., p. 10.

23. Jorge Luis Borges, "Kafka and His Precursors," in Borges, op. cit., p. 201.

24. For a discussion of objectivity as it applies to some institutions other than art, see my "Informing the Public: Ethics, Policy Making, and Objectivity in News Reporting," *Philosophy in Context* 20 (1990): 1–21, also published in my *Philosophy as Diplomacy* (Amherst, N.Y.: Humanity Books, 1994), pp. 40–52.

25. José Ortega y Gasset, "The Dehumanization of Art," in José Ortega y Gasset, *The Dehumanization of Art and Notes on the Novel* (Princeton, N.J.: Princeton University Press, 1948), p. 4.

26. Ibid., pp. 5–6.

27. Ibid., p. 7.

28. Ibid., p. 8.

29. Ibid., p. 9.

30. Ibid., p. 5.

31. Ibid., p. 25.

32. Ibid., p. 11.

33. See my *Philosophy as Diplomacy*, pp. 53–66.

34. See my *Philosophy as Diplomacy*, passim, and the preceding essays.

35. See essay 7 in *Philosophy as Diplomacy*.

Dialogue

— From what you said, I gather that Homer could cease to be a classic.

— You gather correctly, though I think it is highly unlikely that such a thing will happen in the foreseeable future.

— After your earlier discussion of cultural fragmentation and the multiculturalism issue, I am not as sure about Homer's staying power as you are. But something else makes me uneasy: what you said about shared and good reasons. I agree that objectivity entails good reasons. Yet I am uneasy with the fact that different people—say, people from different cultural traditions or with different aesthetic sensibilities—will not always agree on what constitutes a good reason. When this happens, the disagreement may become irreconcilable. This is not a hopeful prospect.

— I agree that such a prospect provides grounds for uneasiness. Yet, it does not undermine objectivity, however much it may undermine the overcoming of cultural or artistic fragmentation. Besides, we should not jump to conclusions. As I argued, the view that certain disagreements about reasons are irreconcilable is merely a hypothesis. Also, even if it is true that they are irreconcilable now, this situation may change later. It all depends on what we do about it.

— But what should we do about it? My uneasiness is not helped by the mere possibility that disagreements just might be overcome.

— It is not a mere possibility. As I have and will further evidence in this book, human history is full of cases in which yesterday's disagreements are overcome by today's common ground. One example is the sort of social settlement of artistic disagreements I discussed in the preceding essay.

— I can see that. Yet, how do we know that there are no blind spots, that all disagreements will be overcome?

— If what you are seeking is certainty, I am sorry to have to disappoint you. Life is risky, our future is uncertain, and a sound theory should both reflect these facts and help us deal with them. This is what the theory I am formulating attempts. I would consider it a failure if it disregarded or misinterpreted these obvious risks and uncertainties. If it did, it would be unrealistic and of little help in dealing with them.

— Fine; but you would at least grant me that some common ground is worth seeking.

— I do.

— How much of a common ground, however, is enough for fragmentation not to have the best of us?

— As I said before, we need enough common ground to keep the dialogue going—not less, but not more than that.

— Yet, more common ground than that would be welcome.

— So long as it does not exclude significant public insights and responses. After all, so long as diversity does not seriously disrupt predominant ecologies, it tends to breed creativity.

— Are you, then, suggesting a path between ecological stability and diversity?

— Actually, I am suggesting that, just as with strictly biological ecologies, social ecologies and the intellectual—in other words, artistic and literary, scientific, and philosophical—ecologies embedded in them often remain stable by remaining adaptable through diversity.

— I noticed you said this is *often* the case. Do you mean to imply it is not always so?

— Of course. Diversity does not help an ecology's adaptation to new situations when, perhaps due to increased external pressures, the diverse components clash to the point of going their own separate ways. Indeed, there is evidence that this is also the case with biological ecosystems. Russell W. Graham and others used a large database from the United States to examine the response of mammals to the warming since the last glaciation and found that biological ties within ecosystems did not prevail under these external pressures. In a period of a few thousand years, ranges of species shifted in different directions and at different rates, hence, coherent mammalian ecosystems were not maintained (Graham et al., *Science*, June 14, 1996, pp. 1601–1606).

— Then how could a minimum of common ground be necessary in social and intellectual ecologies?

— The minimum of common ground I previously discussed is only aimed at helping meaningful dialogue flourish. This may, in fact, undermine prevalent ecologies because, as we previously discussed, there may be good moral reasons—for example, that the prevalent ecologies are objectionable—for favoring ecological change and, to the extent we are likely to be effective, for acting so as to bring about such change.

— So the common ground you mentioned does not necessarily have the function of preserving prevalent ecologies?

— Right. It may help do away with them, *but not recklessly*. For, unless undercurrents in the prevalent ecologies can be brought to the fore and develop into new ecologies, the social conditions for meaningful dialogue about such things as art, literature, science, philosophy, and whatever people value in life would be absent.

— I can see your point, and I can also see how the changes you suggest can occur in society generally, or in artistic or literary communities. But I have difficulty seeing how the same process can take place in science. In it, bodies of belief seem coherently structured, and any changes they undergo seem determined by scientific experiments on a case-by-case basis, not, as in your account, by stands people take concerning these bodies of belief as wholes.

— Granted, science proceeds through a dialogue whose rules are stricter than those of ordinary life, and, at one level, it can be said that science today proceeds according to roughly the same protocols throughout the planet and is unaffected by cultural diversity. Yet, I believe scientific change also displays the general features that appear in other cultural changes.

— How so?

— I will attempt to describe this in the next essay.

— I can't wait.

— I hope I won't disappoint your expectations.

— So do I.

Ecologies of Belief:
Science, Philosophy, and
Ordinary Thinking
in a Fragmented World

BELIEF TENSIONS

Should one, on the side of skepticism and cautiousness, better risk loss of truth than chance of error or, on the side of faith and boldness, better risk chance of error than loss of truth? People are variously divided on this point. Their views range from utter skepticism to total faith. Within this range, some ordinary beliefs conflict with still other ordinary beliefs, while finding support in some and conflicting with still other scientific beliefs. Among the scientific beliefs involved in these conflicts, some are predominant, while others represent undercurrents in the scientific community. At any rate, whether scientific or not, these beliefs are in tension and in interaction with clusters of other beliefs, hence not isolated but constituting networks of interconnected, interactive, and, to some extent, mutually adapted beliefs, or, in other words, *ecologies of belief*. Further, they are embedded in a variety of attitudes, practices, and institutional arrangements that, as discussed in the preceding essays, form *social ecologies*.

In this essay, I will explore ecologies of belief, their mutual and internal tensions, and their interconnections with social ecologies. As I have done in this book so far, my inquiry will be based on a discussion of historical cases together with a critical assessment of the features these cases display. I will ask the following questions: What are some scientific instances of ecologies of belief? What are the functions of these ecologies in scientific inquiry? What, if any, are their functions concerning everyday and philosophical inquiry? What are some nonscientific instances of ecologies of belief? What are their functions concerning everyday and other forms of inquiry?

I will support the following theses. First, ecologies of belief partly regulate scientific inquiry in a generally double-edged way: They tend to conserve predominant scientific theories, while at the same time making room for undercurrents. Second, these undercurrents focus on anomalies within the predominant theories or on conflicts with other theories or with ordinary beliefs, attitudes, or institutions and may, however partially, point to theories under which the anomalies or conflicts would disappear. Third, sometimes, through this interactive process between prevalent beliefs and undercurrents, significantly developed theories are formulated or modified or useful techniques are developed that lead to superseding the previous theories. Fourth, ecologies of belief partly regulate philosophical and everyday thought in a similar, though much less structured, manner.

In supporting these theses, I will begin by examining two historical examples of the influence of ecologies of belief on scientific developments. The first example will focus on the birth of modern mechanics and how it was affected by the ecologies of belief predominant in the scientific communities and, indeed, the entire societies in which this development took place. The second example will study the influence of ecologies of belief at the personal level. It will accordingly focus on a particular scientist, Michael Faraday, and examine his early scientific development in interaction with the predominant ecology of belief in the scientific community of his time, and the social ecology in which it was embedded.

ECOLOGIES OF BELIEF AND SCIENTIFIC DEVELOPMENTS

The Birth of Modern Mechanics

Copernicus's Proposal

Copernicus's book *On the Revolution of the Heavenly Spheres* was published in 1543. More than fifty years later, at the beginning of the seventeenth century, it was still to be seen whether there would be a Copernican revolution.[1] Indeed, Copernicus himself had only proposed limited changes in planetary theory: from a geostatic to a heliostatic description of the movements of the components of the solar system.[2]

Why should the Copernican system have been preferred? This question concerns the ecologies of belief within which the said preference had to be justified at Copernicus's time. They included two main types of components: scientific beliefs and popular beliefs. Let us first consider the Copernican system from the standpoint of the predominant scientific beliefs of the time.

The Copernican System and the Predominant Beliefs of Physicists at the Beginning of the Seventeenth Century

The Copernican system had some features that made it appealing to scientists. First, it easily explained the retrograde motion of planets and showed why their positions in relation to the sun led to this retrograde motion. Each planet had a different period of orbital revolution around the sun. The further a planet was from the sun, the greater the period.[3]

Second, the Copernican system provided a way of computing the distances of the planets from earth and from the sun. Consider, for example, Venus. Observations made each night evidence when Venus can be seen at its farthest elongation from the sun. When this occurs, a line from earth to Venus, EV, tangent to Venus's orbit, is perpendicular to a line from the sun to Venus, VS. From simple trigonometry, it follows that VS/ES = sine α. In the Copernican system, the distance ES—the average size of the radius of the Earth's orbit—is called an "astronomical unit," 1AU, and makes it possible to rewrite the previous equation as follows: V-S = (sine α) $\times$ 1AU. This allowed Copernicus, quite accurately, to determine the planetary distances in astronomical units.[4]

Simplicity is sometimes said to be one of the advantages of the Copernican system. This, however, is debatable. For this system turned out to be more complicated than involving, as it is sometimes suggested, merely a single circle for each planet, a single circle for the moon, and two different motions—rotation and orbital revolution—for each planet and the moon. Let us see why.

Copernicus aimed at describing planetary orbits around the sun that led to predictions conforming to actual observation. The above simplistic system, however, did not quite agree with observation. Its accuracy partly depended on the uniform motion of the planets around the sun. But Copernicus thought its predicting accurately to be impossible with the sun actually at the center. Hence, he placed the sun at some distance off this point and the center of the solar system became the center of the earth's orbit or, as I. Bernard Cohen calls it, a "mean sun," and the system became heliostatic rather than heliocentric.[5]

Further, to ensure the system's accuracy, and given that the planets did not simply move in circular orbits, Copernicus had to use circles moving on circles, just as Ptolemy had done. The main difference was that Ptolemy had used the circles to explain retrograde motion too, while Copernicus used them only to predict their positions accurately.[6]

As for the disadvantages of the Copernican system, the most notable one is the fact that the system's description of planetary motions did not fit the then-current Aristotelian dynamics used to explain motions in the solar system—a dynamics Copernicus never explicitly rejected. Yet, if earth was just another planet, as Copernicus indicated, the then-predominant Aristotelian view that earth and the other planets were made of different materials, obeyed different physical laws, and, therefore, behaved differently, could not be true. Also, if the planets traveled around the sun, something for which predominant Aristotelian physics had no place kept them from moving away from the sun, catapulted into space. And something—also extraneous to predominant Aristotelian physics—kept the moon from getting lost.[7]

These were not minor points. For they presented the predominant Aristotelian physicists with the following scientific choice: Either to adopt the Copernican description of the motions of solar system components and end up with a physical theory incapable of explaining these and any other motions, or to stick to Ptolemy's description and retain a complete—however cumbersome—physical theory. They generally chose the latter, which went well with a network of nonscientific beliefs interconnected with the previous ones in the predominant ecology of belief. Let us examine these next.

The Copernican System and Predominant Beliefs and Practices in Europe at the Beginning of the Seventeenth Century

Besides its scientific component, the predominant ecology of belief in Europe at the beginning of the seventeenth century included, most notably, a largely Judeo-Christian philosophical tradition—a philosophical ecology or network of interconnected philosophical beliefs in itself—and was interconnected with the predominant social ecology of practices and institutions whether of Catholic or Protestant persuasion.

People generally believed that the earth was unique and the center of the universe, which gave humans a prominent place in the entire order of things. This belief was undermined by the Copernican system, where the earth was just another planet, which left open the possibility that there, perhaps, were other planets inhabited by beings comparable, if not superior, to humans.

By and large, Europeans were simply unprepared for this line of thought. Besides, their senses told them that the earth was not just another planet. For, to them, the visible planets looked bright, like stars, while the earth did not.

And, if the senses were not enough, the Bible came to the rescue by repeatedly mentioning a moving sun and a fixed earth. Indeed, having heard of Copernicus's ideas before the publication of Copernicus's work, Luther condemned them on the grounds that they conflicted with the Bible.[8]

Late Sixteenth- and Early Seventeenth-Century Undercurrents

Like all ecologies, the social ecology and its component, the ecology of belief, predominant in Europe in the late sixteenth century were not free from undercurrents. With regard to the place of the earth in the solar system, however, these undercurrents were well in the minority. Indeed, only the Pythagorean-Hermetic tradition accepted the heliocentric view. It did so on the grounds that it was only fitting for the sun to be at the center.[9]

There is evidence that this tradition influenced Copernicus during his years of study at the universities of Bologna and Padua. At Bologna, his associate and teacher, Domenico Maria de Novara, knew the Florentine Neoplatonists and had translated Proclus and Hermes Trismegistus. And the influence of these thinkers can be found in Copernicus's writings:

> In the middle of all sits the Sun enthroned. In this most beautiful temple could we place the luminary in any better position from which he can illuminate the whole at once? He is rightly called the Lamp, the Mind, the Ruler of the Universe; Hermes Trismegistus names him the visible God, Sophocles Electra calls him the All-seeing. So the sun sits as upon a royal throne ruling his children the planets which circle around him.[10]

Neoplatonism and Hermetism aside, a number of established astronomers read Copernicus's work and took it seriously. Yet, they did not consider it part of established astronomy. These included Robert Recorde, who taught mathematics in London, and Michael Maestlin, a professor of astronomy at Tübingen and Kepler's teacher. Indeed, in 1596, Maestlin oversaw the publication of Kepler's first book, *Cosmographic Mystery*, and appended Rheticus's 1540 *First Narration* (one of the works where Copernicus's ideas had been initially mentioned) with a preface in praise of Copernicus. And, outside astronomy's circles, radical thinkers, typically anti-Aristotelians, many of them nonscientists, found the Copernican account very appealing.[11]

Kepler and Galileo

In his first work, Kepler, also influenced by Neoplatonism, assumed the view that the universe is constructed according to geometric principles. But he also believed that, besides mathematics, astronomy had to rest on sound physical principles. Further, he believed that the crystalline spheres supposed to constitute the *immutable* sky in traditional astronomy did not exist. For, if they did, the new star of 1572 and the comet of 1577, which made the sky something other than immutable, would not have been observed. In addition, Kepler was convinced that celestial motions could be explained by the same principles that explained the motions of terrestrial objects. But he still adhered to Aristotelian dynamics. Hence, in his *New Astronomy Founded on Causes, or Celestial Physics Expounded in a Commentary on the Movements of Mars*, published in 1609, he attempted to employ the basic Aristotelian principles of dynamics to explain the motions of the bodies beyond the moon.[12]

In doing this, Kepler was true to the predominant scientific beliefs of the time. He assumed the physicists' beliefs on dynamics and tried to apply them to the objects beyond the moon. But the simplicity thus gained by the

Copernican-Keplerian theory still faced problems. First, from the standpoint of astronomy, even after Galileo's invention of the telescope and his turning it to the sky in 1609, no stellar parallax—in other words, no change in the position of the fixed stars as earth was supposed to travel around the sun—was observed. Second, from a common sense standpoint, people found it hard to imagine that, despite what they saw every day, the sun did not rise or set and earth did not stand still. These and other problems concerning motion had to be resolved before the new astronomy was accepted.[13]

Nonetheless, at the beginning of the seventeenth century, the proponents of the Copernican system were not an insignificant minority any longer. This is not to say that astronomers had become Copernican in great numbers—though there were more Copernican astronomers than in the 1540s. Nor is it to say that new data supported the Copernican system. Indeed, the body of data gathered by the greatest practical astronomer of the time, Tycho Brahe, provided no direct support for Copernicanism, and Tycho Brahe himself was an anti-Copernican. Kepler's work was necessary, but by no means sufficient to use this body of data in support of the Copernican system.[14]

The change concerning the Copernican system took place chiefly in society at large, where this system had become familiar and its supporters numerous, especially among mystics. As a result, the ecclesiastics' position stiffened. At the University of Pisa, where Galileo had studied and taught for a short while before going, in 1592, to the more liberal University of Padua, the scholastic philosophers declared his opinions to be false and contrary to Aristotle's authority. In 1615, Galileo was summoned before the Roman Inquisition and was made to abjure the Copernican system. In 1616, the work of Copernicus was placed on the Index of forbidden books, where it was kept until 1835.[15]

Galileo, however, did not change his opinions. Indeed, in 1632, with the permission of Florence's inquisitors, he published his *Dialogue Concerning the Chief Two World Systems*. In this book, he addressed the problems still facing the Copernican system by focusing on the problem of vertical fall: How, if the earth traveled through space, could a ball keep up with the tower from which it was dropped? Galileo provided the answer by appeal to a notion akin to that of inertia, thus beginning to provide the dynamics in a new system of mechanics for objects under the moon. He wrote: "Keeping up with the earth is the primordial and eternal motion ineradicably and insepa-rably participated in by this ball as a terrestrial object, which it has by its very nature and will possess forever."[16]

Even though this conception helped develop a new dynamics, Galileo did not entirely break with tradition. Concerning celestial motions, he stayed close to the old astronomy in holding that circular, and only circular, motion was compatible with an ordered universe. That is, about twenty years after the publication of Kepler's *New Astronomy*, Galileo, who also intended to support the Copernican system, ignored the problems addressed by Kepler's celestial mechanics, which had led Kepler to describe celestial motions by means of elliptical orbits and make the need for circles, and circles of circles, obsolete.[17]

Just as Kepler's, Galileo's case illustrates how new views are developed in general accordance with, but taking advantage of the tensions within ecologies of belief and the social ecologies within which these are embedded. Eventually, the work of these astronomers received a unified formulation in Sir Isaac Newton's *Principia*, published in 1687, but, like the work of his predecessors, it was not accepted until later, once it was found to fit the current ecology of belief and social ecology.[18]

IMPLICATIONS OF THE PREVIOUS CASE FOR THIS ESSAY'S THESES

The case of the birth of modern mechanics just discussed provides some confirmation for this essay's theses. First, ecologies of belief partly regulate scientific inquiry in a generally double-edged way: They tend to conserve predominant scientific theories, while at the same time making room for undercurrents. Second, these undercurrents focus on anomalies within the predominant theories or on conflicts with other theories or with ordinary beliefs, attitudes, or institutions and may, however partially, point to theories under which the anomalies or conflicts would disappear. Third, sometimes, through this interactive process between prevalent beliefs and undercurrents, significantly developed theories are formulated or modified or useful techniques are developed that lead to superseding the previous theories. Fourth, ecologies of belief partly regulate philosophical and everyday thought in a similar, though much less structured, manner.

One might think, however, that the type of situation exemplified by the birth of modern mechanics was peculiar to other times and has nothing to do with the development of science in the modern, let alone the contemporary, world, where scientific inquiry, at least pure scientific inquiry, has become largely isolated from social influences. This however is false. For, first, whatever isolation science has attained is a double-edged sword. People may leave scientific research alone so long as the paths of science and ordinary life do not cross. Yet, when they do cross, science's isolation undermines it, at least in that the general public as well as legislative and executive bodies are far from supporting science budgets even in countries that have traditionally supported scientific research. The legislative history of science budgets in the United States during the early 1990s makes this plain.[19]

Second, the role of ecologies of belief is often crucial even from the standpoint of science's internal development and, indeed, from the standpoint of the development of particular scientists. I will next provide evidence for these points through a brief examination of scientific and personal development through the initial stages of Michael Faraday's groundbreaking work in field theory—an area significantly distant from everyday social concerns.

Michael Faraday's Early Scientific Development

Newtonian Background

The publication of Isaac Newton's *Principia* in 1687 had brought order into the world of physical objects, from falling stones to planets. What remained to be done was to extend Newton's theory to the entire world of physics, including light, heat, electricity, and magnetism. Toward the end of the eighteenth century, all these phenomena had been reduced to imponderable fluids that obeyed the laws of Newtonian mechanics. Concerning electricity and magnetism, for example, it was generally agreed that all phenomena falling under this category could be explained only by the assumption of imponderable fluids exercising action at a distance on each other.[20]

To be sure, unanimity was not part of the picture at the turn of the century. Sir Humphry Davy, for example, raised doubts about the doctrine of the materiality of heat in *An Essay on Heat, Light and the Combinations of Light* (1799). Indeed, once his fame and self-confidence grew, Davy rejected the fluid theory of electricity, and, since 1807, he constantly referred to electric powers and electric energies instead. Davy, however, was in the minority, and, in 1815, his views were sharply criticized in the *Annals of Philosophy*.[21]

Faraday's Upbrining

Michael Faraday was born in Newington in Surrey, England, on September 22, 1791. Until a few years before his birth, his family had lived in Outhgill, where his father's income as a blacksmith and his mother's income as a maid-servant at a nearby farm had sufficed to take care of the family. But the 1789 French Revolution led to a gradual depression in England, Faraday's father's income as a blacksmith suffered, his health began to fail, and, hearing of better prospects around London, he moved with his family to Newington shortly before their son Michael was born.

During those early years, the family lived in cramped quarters over a coach-house. Food was scarce. Michael Faraday's elementary education was almost nil. But his parents were members of the Sandemanian Church, whose emphasis on serenity, kindness, asceticism, and a sense of belonging to the Sandemanian community, and the freedom from nonmembers' concerns with such things as guilt and fear of the future helped them to cope with the circumstances. Indeed, these traits were going to be characteristic of Michael Faraday for the rest of his life.[22]

He began to work as an errand boy for Mr. Riebau, a kind and generous bookseller and bookbinder, at the age of thirteen. During his years there, he read a wide variety of books. He developed a keen interest in electricity, and, regarding matters of method, he was significantly influenced by Dr. Isaac Watts's *The Improvement of the Mind*. In this volume, Watts advised young people to keep a notebook, attend lectures, exchange correspondence with a person of similar interests and attainments, and take part in a small discussion group. Faraday did all these things. He also took very seriously Watts's Lockian emphasis on the dangers of imprecise language, the importance of observation, and the shortcomings of hasty generalization. These views helped place Faraday in a critical tradition that, no doubt, was commonplace in the nineteenth-century learned communities but was hardly accessible to the young uneducated Faraday.[23] At any rate, they were going to guide him well into his mature years.

In 1812, Faraday felt that he was at a crossroads of his scientific development. He had religiously attended the meetings of the City Philosophical Society since its inception in 1808; relentlessly read on science; taken private writing and drawing lessons for years; regularly exchanged correspondence and met with the well-educated Benjamin Abbott to discuss chemical problems of common interest; set up a small laboratory in the back of Mr. Riebau's shop; and also, with Mr. Riebau's help, gained entry to the Royal Institution, where he would attend four lectures by Sir Humphry Davy. Yet, he had no access to the proper scientific facilities.[24]

Hence, he applied for whatever job, however menial, he could be given at the Royal Institution. His request was ignored by the institution's president; but, probably through a recommendation from Mr. Dance, one of Mr. Riebau's clients whom Mr. Riebau had shown some of Faraday's writings, Faraday was temporarily hired as an amanuensis for Sir Humphry Davy. This, together with the help of a variety of coincidental events, most notably, the summary discharge, as a result of a brawl, of William Payne, who had served in the laboratory of the Royal Institution, led to his becoming assistant to Davy in this laboratory, a Royal Institution's helper, and, eventually, an amanuensis to Davy in his 1813 trip to Napoleon's France and later to Italy.[25]

After about two years, Davy decided to return to England. During the trip, Napoleon had been imprisoned in Elba. Toward the trip's end, he escaped, threatening Europe again. This apparently was part of Davy's motivation for returning home. At any rate, Faraday also went back home.

Through his rich interactions with foreign peoples and places in Europe, Faraday had changed a lot. He had become keenly aware of cultural differences, had learned much French and Italian, had met some of the chief scientists of the time, and had gained acceptance in high levels of society. Yet, he still was motivated by the pursuit of scientific truth, did not hold high society in high esteem, and did not much care for the political rumblings caused by Napoleon's return or by the—during his life, inveterate—conflicts between England and France. International politics and, as his later activities in the British scientific community made plain, any politics left him entirely unmoved.[26] Indeed, upon his return, in a manner faithful to his Sandemanian upbringing, he distanced himself from the affairs of the world at large and ascetically immersed himself in his research.

Faraday's Experimental Beginnings

Michael Faraday's work as a young amanuensis helped Davy rewrite his *Elements of Chemical Philosophy* in Paris in 1814. This experience helped Faraday flourish. His early scientific development, however, took place against the conflictive theoretical background between Davy's rejection of the fluid theory of electricity and the general acceptance of this theory.

No wonder that, between 1816 and 1819, when Faraday delivered a series of lectures to the members of the City Philosophical Society, he seemed to be of many minds concerning the nature of electricity and magnetism. In his notes to the introductory lecture, for example, he claimed the nature of electricity to be unknown. During the lectures, however, he shifted between the conception of electricity as a power and the fluid theory. His treatment of magnetism showed the same unsettledness; be seemed quite uncertain of what to believe on the subject. But even when, toward the end of the three-year period of his lectures, his theoretical preferences began to lean toward the immaterial nature of electricity, magnetism, heat, and light, he saw these views as mere preferences.[27] Faraday approached his research with a strongly experimentalist, largely antitheoretical bent. In his May 26, 1819, lecture, for example, he referred to theoretical differences as "merely squabbles about words."[28]

The Discovery of Electromagnetic Induction

This attitude about theories could still be found three years later, in his "Historical Sketch of Electromagnetism," where Faraday was still suspending judgment on theoretical matters. Yet, in an undeveloped manner, he began to mention the notion of an intermolecular state he was to pursue later:

> Whatever the cause which is active within the connecting wire, whether it be the passage of matter through it, or the induction of a particular state of its parts, it produces certain very extraordinary effects.[29]

The study of the history of research on electromagnetism, however, significantly propelled Faraday's work on electricity and magnetism. Further, his stand on electromagnetic theory was substantially strengthened by his discovery of electromagnetic rotation toward the end of his study of the history of electromagnetism. His results appeared in "On Some New Electro-Magnetical Motions, and the Theory of Magnetism," raising questions about every theory of magnetism proposed until then.[30] Faraday still was holding back on what theory to adopt; but he did express serious reservations about existing theories, especially Ampère's, the most widely accepted one. His crucial disagreement with Ampère concerned the nature of the interaction between electricity and magnetism. For Ampère, it was a straightforward attraction or repulsion between moving electrical currents. For Faraday, it was a more complicated affair where the basic force was the circular force that pushed a magnet around a current-carrying wire or vice versa.[31] Electrical and magnetic phenomena were thus reduced to the circular force—however partial, a theoretical result no doubt.

The Wollaston Affair

The highly respected English scientist William Hyde Wollaston had talked of the rotation of a current-carrying wire about its own axis; but he had never published these views. In contrast, Faraday's views pointed to the rotation of a current-carrying wire about the pole of a magnetic needle. In addition, Faraday applied his discovery in devising the first electric motor, and he announced his discovery in the October 1821 issue of the *Quarterly Journal of Science*, which almost immediately made him famous throughout Europe. Yet, nowhere in this publication did Faraday make any reference to Wollaston.

He claimed that he had gone to Wollaston's house to ask for permission to refer to his views, but Wollaston was out of town, and hence he had proceeded to publish his article without referring to unpublished views he had no right to mention. He expressed regret not to have delayed publication until Wollaston's return. But these words did little to curb rumors of plagiarism in the scientific community. After all, Wollaston had talked of rotations, and the distinction between rotation around an axis and rotation around a magnetic pole was too fine for many to see the difference.[32] The affair eventually died out, but, as we shall see, it had the lasting effect of reinforcing Faraday's cautiousness in making his scientific opinions public.

In Search of a New Theory

Faraday also realized that his discovery of electromagnetic rotation called for a modification in his idea of the state of a current-carrying wire. In order to cause a dynamic rotation, it had to be partly dynamic, not merely a static arrangement of particles. But he had distanced himself from the theory of fluids. He had to explain the transmission of force appealing neither to the direct impact of particles nor to action at a distance. Eventually, he found his sought candidate in the undulatory theory of light, which had recently been proposed by Fresnel, whose work was officially recognized by the British scientific community in 1827 and translated between 1827 and 1829.[33]

Why, however, should electricity and magnetism, as Fresnel had proposed for light, involve a wave motion? Faraday, who loved music and had begun dabbling into the study of sound, found logical support for his extrapolation in John Herschel's 1830 *A Preliminary Discourse on the Study of Natural Philosophy*, in which the analogy between light and sound was strongly emphasized. On the strength of arguments by analogy, Faraday hoped to use experimental results in acoustics to clarify the nature of vibrations in general.[34] This effort proved fruitful in August 1831 with his discovery of electromagnetic induction.

Field Theory and the Predominant Ecology of Belief in Early Nineteenth-Century Science

Faraday's efforts also led, in March 1832, to his taking a definite, though extremely cautious, stand in favor of a generalized version of the vibratory theory. He formulated his views, which clashed with the prevalently accepted ones—those of Ampère—and submitted them in a sealed note to the Royal Society.[35]

This approach seemed to have three purposes. First, to give time for Faraday's developing his ideas experimentally. Second, to avoid a mishap similar to that which had caused the Wollaston affair a decade earlier. Third, to avoid controversy with those—certainly more numerous—scientists who upheld opposite theories at the time. Yet, in his correspondence with Whewell, Faraday hypothesized the existence of the electrotonic state to explain the creation of current when the current in the primary ceased.[36] Indeed, he later abandoned the notion of an electrotonic state in favor of a *field of force*.[37] Through the various interactions I have described, Faraday's work had become significantly theoretical, a feature it was to retain in the next decades.

Various discoveries still awaited Faraday along his way toward the formulation of field theory. In 1837, he discovered that electrostatic induction took place along curved, not straight, lines, which implied that electrostatic induction was not action at a distance. In the late 1840s, he discovered the diamagnetism of gases. In 1850, he explored the connection between the magnetic properties of oxygen and variations in terrestrial magnetism. All along this thirty-year period, however, his research focused on two questions:

How are magnetic and electric forces transmitted through space? And how are these forces related to ponderable matter? Modern field theory was to emerge from the answers given to these questions by Michael Faraday, James Clerk Maxwell, and Albert Einstein.[38]

Implications of the Cases So Far Discussed for This Essay's Theses

Faraday's case helps build on the birth of modern mechanics previously discussed. For it provides additional confirmation for our first three theses, showing them applicable not just during the scientific revolution but also in modern science. First, ecologies of belief partly regulate scientific inquiry in a generally double-edged way: They tend to conserve predominant scientific theories, while at the same time making room for undercurrents. Second, these undercurrents focus on anomalies within the predominant theories or on conflicts with other theories or with ordinary beliefs, attitudes, or institutions and may, however partially, point to theories under which the anomalies or conflicts would disappear. Third, sometimes, through this interactive process between prevalent beliefs and undercurrents, significantly developed theories are formulated or modified or useful techniques are developed that lead to superseding the previous theories.

Further, as the language of scientists and others in the works so far discussed makes plain, scientific inquiry traditionally, and into modern times, was hardly distinguishable from philosophical inquiry. Hence, whenever ecologies of belief affected scientific developments, they also affected philosophical developments in the modern world. Indeed, this situation has continued in later times, despite the fact that science and philosophy have taken different paths. For philosophy still takes upon itself to resolve the tensions between scientific and everyday thought.

This leads to the manner in which ecologies of belief, through both their scientific and ordinary components, affect ordinary thought. For, in the modern and contemporary world, ordinary thought has been increasingly exposed to scientific developments—say, to evolutionary theory and contemporary cosmological theories—and to engineering and technology applications of science—from computers, through the green revolution, to genetic engineering. As a result, it has been sensitized to the conflicts that develop between scientific and technological developments, on the one hand, and traditional everyday beliefs—say, that humans are not mere machines but are persons and have dignity—on the other. Through the largely unstructured interactions between its various scientific and everyday components, ecologies of belief regulate not just scientific changes, as previously argued, but also changes in everyday thought. Given all this, we have reasons in support of our fourth thesis concerning not just the scientific revolution but the modern and contemporary world: Ecologies of belief partly regulate philosophical and everyday thought in a similar, though much less structured, manner.

This Essay's Theses and Contemporary Trends in the History of Science

It should be noted that the theses just formulated are not general statements about the essential nature of science or the invariable genesis of scientific revolutions. Theses of a more general sort have been advanced, for example, by Sir Karl Popper, who sees science as constantly on the verge of revolution and bases this view on a rationalization of what scientists do.[39] Equally general theses have been advanced by Thomas Kuhn, who has a less volatile view of science and claims to base this view on examples of what scientists actually do.[40] As L. Pearce Williams has stated, however, the cases used to support these very general theses are often too few. And as our example of Michael Faraday makes plain, scientists sometimes have a publicly conservative, but privately radical, scientific agenda.[41] Which view of scientific development, a Popperian or a Kuhnean one, do such cases support?

By contrast, the theses in this essay are more modest and make room for more complex processes in scientific development. They focus on ecologies of both scientific and nonscientific belief, where these beliefs variously affect scientific inquiry in both the individual and the scientific community and at various levels of these, and how this interplay within ecologies of belief affects philosophical inquiry and everyday thought. The theses further state that these ecologies of belief partly—not entirely—regulate inquiry, which sometimes leads to the formulation and even adoption of new scientific views, and that together with scientific developments, they also, though in a more unstructured way, regulate changes in philosophical and everyday thought.

Such a more modest stand is better supported by a few cases than the more general positions advanced by Popper and Kuhn. It also exemplifies the kind of work that should be carried out in a very general fashion in order to be in a position to assess the relative merits of more general views such as those of Popper or Kuhn.

NOTES

1. See, for example, Richard S. Westfall, *The Construction of Modern Science: Mechanisms and Mechanics* (New York and London: John Wiley and Sons, 1971), p. 3.

2. Ibid. For the reasons to say *heliostatic* rather than *heliocentric* and for a characteriza-

tion of the changes proposed by Copernicus, see I. Bernard Cohen, *The Birth of a New Physics* (Garden City, N.Y.: Anchor Books, 1960), pp. 48–57.

3. I. Bernard Cohen, op. cit., pp. 49, 57.

4. For a more detailed discussion of this point, see ibid., pp. 53–54.

5. Ibid., p. 56.

6. Ibid., pp. 56–57.

7. Ibid., pp. 59–61.

8. Ibid., pp. 61–63. Copernicus's theory had been known in certain circles for many years before the 1543 publication of his *De revolutionibus orbium caelestium*. It had been known through his 1512 *Commentariolus*, through rumor, and through Rheticus's 1540 *Narratio Prima*. See, for example, Marie Boas, *The Scientific Renaissance: 1450–1630* (New York: Harper & Row, 1962), p. 91.

9. Hugh Kearney, *Science and Change: 1500–1700* (New York and Toronto: McGraw-Hill, 1971), p. 104.

10. Quoted in ibid., pp. 98–100. It can also be found in Thomas S. Kuhn, *The Copernican Revolution* (Cambridge, Mass.: Harvard University Press, 1951), p. 128.

11. Boas, op. cit., pp. 94–97.

12. Westfall, op. cit., pp. 3–13, especially pp. 4–5.

13. Ibid., pp. 13–15.

14. Boas, op. cit., p. 127.

15. Stephen F. Mason, *A History of the Sciences* (New York: Collier, 1962), pp. 153, 160–61.

16. Quoted in Westfall, op. cit., p. 17. See also the discussion on pp. 17–18.

17. Ibid., pp. 9, 18; and Mason, op. cit., pp. 161–64.

18. Kearney, op. cit., pp. 187–96.

19. For the variety of controversies surrounding the applications of science today, see my *Contemporary Moral Controversies in Technology* (New York and Oxford: Oxford University Press, 1987). For science budgets in the early 1990s, see the relevant issues of *Science* for the same period.

20. L. Pearce Williams, *Michael Faraday* (New York: Simon & Schuster, 1971), pp. 54–55.

21. Ibid., pp. 66, 69, 70.

22. Ibid., pp. 3–8. See also Geoffrey Cantor, *Michael Faraday, Sandemanian and Scientist: A Study of Science and Religion in the Nineteenth Century* (London: Macmillan; New York: St. Martin's Press, 1991), passim.

23. Ibid., pp. 10–13.

24. Ibid., pp. 15–28.

25. Ibid., pp. 28–31.

26. Ibid., pp. 41, 348–59.

27. Michael, Faraday, "Forms of Matter," Institution of Electrical Engineers chemistry lecture, May 26, 1819, quoted in Williams, op. cit., pp. 85–86.

28. Quoted in Williams. op. cit., p. 86.

29. Michael Faraday, "Historical Sketch of Electromagnetism," *Annals of Philosophy* 2 (1821): 197; Williams, op. cit., p. 154.

30. Michael Faraday, "On Some New Electro-Magnetical Motions, and on the Theory of Magnetism," *Quarterly Journal of Science* (October 1821); Williams, op. cit., p. 157.

31. Faraday, "On Some New Electro-Magnetical Motions, and on the Theory of Magnetism," as it appeared in *Experimental Researches* 2 (1859): 132; Williams, op. cit., p. 161.

32. Williams, op. cit., pp. 156–61.

33. Faraday to Eilhard Mitscherlich, *Royal Institution, letter*, August 4, 1830, in Williams, op. cit., pp. 175–77.

34. Williams, op. cit., pp. 177–79.

35. Royal Society Manuscripts, Williams, op. cit., p. 181.

36. Faraday to Whewell, Royal Institution, May 5, 1834, and May 20, 1840, cited in Williams, op. cit., p. 198.

37. Williams, op. cit., pp. 202–206.

38. Ibid., pp. 205–206, 296, 394–99.

39. See, for example, Karl Popper, *The Logic of Scientific Discovery* (New York: Scientific Editions, 1961) and *Conjectures and Refutations* (New York: Harper, 1965).

40. See Thomas S. Kuhn, *The Structure of Scientific Revolutions*, 2d ed. (Chicago: University of Chicago Press, 1970) and later works.

41. L. Pearce Williams, "Normal Science, Scientific Revolutions and the History of Science," in Imre Lakatos and Alan Musgrave, *Criticism and the Growth of Knowledge* (London: Cambridge University Press, 1970), pp. 49–50.

Dialogue

— I think you have done a disservice to science and technology.

— Why?

— Because, given how you conceive of ecologies of belief, these include prevalent ordinary beliefs that contribute to regulate scientific development and can be, and sometimes are, quite arbitrary and irrational.

— No doubt, the influences you mention play an important role in constraining such things as funding for research, when not constraining the very legality of engaging in certain types of scientific inquiry. Think, for example, of the limitations placed on fetal research in the United States during the early 1990s on the grounds that it encouraged women to have abortions. Restrictions such as these constitute a problem for scientific and technological advancement. In our case, they arguably slowed down advances concerning the treatment of Alzheimer's and other diseases. But they need not affect the grounds or methods leading to the acceptance of scientific views by a substantial number of recognized members of the scientific community. In fact, the scientific protocols used today are largely invariant throughout the planet.

— Once again, I think you are more optimistic than circumstances warrant. It seems to me that a significant number of the Third Reich's scientific establishment made it a practice to accept hypotheses concerning Aryan superiority by appeal to arbitrary and irrational beliefs.

— I did not say such aberrant behavior never takes place. There are, after all, some bad scientists. Indeed, pressured by extreme political circumstances like those in the Third Reich, entire groups of scientists can be led to employ questionable scientific standards. My point is rather that this type of influence on scientific conduct need not—and typically does not—affect the adoption of scientific views.

— I am not so sure of that either. Think of the many fabrications of evidence that have occurred in the United States, at least since the 1980s, often prompted by the combination of greed and fear that motivates people to try to secure funding, academic positions, and prestige at any cost.

— Your example is clearly an example of scientific misconduct, not an example of scientific views accepted on arbitrary or irrational grounds. The scientists engaging in such misconduct do not accept the views they claim to have established. They simply try to con others into accepting them. Mind you, they do so by concocting phony scientific evidence, not by appeal to arbitrary or irrational views. Indeed, I wonder how much of such opportunism was driving the claims of the Third Reich's scientists you mentioned instead of their basing their views on ordinary beliefs that had nothing to do with evidence.

— Maybe some were opportunists. But were they all? At any rate, the evidence suggests some were not, while your suspicions are based on sheer speculation. Besides, even in the less dramatic cases you discussed in your essay, scientific views were accepted partly on the basis of ordinary beliefs.

— How so?

— Copernicus, for one, accepted his heliostatic conception of the solar system partly on the basis of Hermetic—that is, religious—beliefs.

— This belief influenced his views, but it was not instrumental in bringing about what is crucial: the acceptance of these views by a substantial number of recognized members of the scientific community. Once Newton's work appeared, modern mechanics was generally accepted, but on the grounds I previously indicated: predictability, completeness, unity, explanatory power, and the like. Nowhere were religious beliefs influential in prompting the acceptance of the new mechanics among a substantial number of recognized scientists, who were more critical and self-critical than that.

— Your position may be defensible concerning such sciences as physics and chemistry; but I believe it is unwarranted otherwise. What about the social studies? Hasn't ideology infiltrated research at least in sociology and political science?

— To be sure, social studies are more liable to be influenced by ordinary beliefs than other branches of inquiry, simply because they are closer to the concerns of ordinary people. But we must be careful not to jump to conclusions. In what way have arbitrary, ordinary beliefs significantly led to the acceptance of hypotheses or theories by a substantial number of recognized members of the social studies community?

— At least some hypotheses and theories have been generally ruled out by appeal to self-indulgence concerning one's own political ideals, when not political correctness.

— What do you mean by self-indulgence concerning one's own political ideals?

— I will give you an example. In foreign-policy studies, the view that democracies do not wage war against each other is so widespread that hardly anyone questions it. This is self-indulgence, and concerns the democratic political ideal of those involved in such studies. Only recently have some scholars begun to pay closer attention to such things as the Fashoda crisis (the 1898 showdown in Sudan between two undisputed democracies—England and France) or the fact that peace among democracies has been statistically significant largely during the time the Soviet Union posed a threat to them—since World War II (*The Chronicle of Higher Education*, July 5, 1996, pp. A6, A7, A12).

— As you indicated, however, scholars are raising questions about this view, some indicating that balance-of-power explanations or, alternatively, wealth-of-democracies explanations provide a better account of the facts. On the other hand, scholars who defend the democratic-peace view do not rely on ideological grounds to adopt it. They insist that democracy teaches peaceful habits, hence, contributes to explain peace among democracies. The same can be said concerning views supposed to be held merely on political correctness grounds. Hence, all you have established is that people may be blinded by ideology into adopting views they believe scientific, but scientific inquiry can, and eventually does, engage in critical scrutiny of such views. What is wrong with science or with my account of ecologies of belief in this regard?

— That people can fall prey to their intellectual limitations.

— This is not science's shortcoming but people's shortcoming. At any rate, within a variety of ecologies of belief, scientific inquiry can and often does find a way to engage in critical scrutiny of views held uncritically by scientists or others.

— Still, ideology can influence the acceptance of scientific hypotheses and theories.

— As I said before, we must be careful not to jump to conclusions. For example, the term *ideology* has many meanings. In which sense are you using it when making suggestions like the one you just made? We need to clarify this, to establish whether this sense is suitable in the context of discussion, and, once a suitable sense of ideology has been characterized, to establish whether, in this sense, ideology is bound to have the effects you suggest on scientific inquiry.

— Yes, we do.

— I am glad you agree, because I will address these and related matters in the next essay.

PART 4

Social Ecologies, Ecologies of Belief, and Philosophy as Diplomacy in an Ideology-Laden World

BETWEEN KNOWLEDGE AND IDEOLOGY: THE DUBIOUS PREDICAMENT OF ECOLOGIES OF BELIEF

The English biologist T. H. Huxley said that science commits suicide when it adopts a creed.[1] This applies not just to science but to every other inquiry, certainly to philosophy, where the search for truth is crucial. As a result, a cluster of poignant questions arises regarding ecologies of belief—as previously explained in this book, regarding networks of interconnected, interactive, and, to some extent, mutually adapted beliefs. Do ecologies of belief commit such inquiries as science and philosophy—which, as argued in previous essays, are carried out within the scope and under the constraints of ecologies of belief—to the adoption of any creed? Are ecologies of belief ideologies? Can scientific, philosophical, or any other sort of inquiry carried out within their scope lead to knowledge, wisdom, and, for that matter, to legitimate decisions and institutions? Or is any such inquiry bound to be a merely arbitrary activity issuing in arbitrary beliefs, attitudes, practices, and social arrangements? How, if at all, can such branches of inquiry proceed along nonarbitrary lines? And, given the nature and functions of ecologies of belief and of ideologies, how should we proceed in forming and transforming beliefs and the conditions that accompany them? These are the main questions I will address in this essay.

I will put forth the following theses: First, for the present inquiry's purposes, ideology is most usefully understood as a network of beliefs, attitudes, conventions, practices, and institutional priorities associated with different, sometimes mutually conflictive, distributions of power and resources. Second, some ideologies may make room for and encourage meaningful dialogue and fruitful interaction in the process of social change. Third, ideologies are not identical or equivalent to ecologies of belief or, for that matter, to social ecologies. Fourth, the transformation of social ecologies in which ecologies of belief are embedded and ideologies involved should proceed

170

along the pragmatic and open-ended lines of a philosophy as diplomacy. Fifth, this philosophy as diplomacy approach is suitable for soundly dealing with people's views concerning changes in ecologies of belief.

In the process of establishing the latter two theses, I will outline an approach for evaluating envisioned alternative situations and proceeding accordingly. This approach will amount to an extension of views I developed in *Philosophy as Diplomacy* to areas of human activity that go beyond policy and decision making and range from linguistic and convention settling in dealing with new situations (partly created, say, by new technologies) to the institution of new social arrangements concerning such things as child rearing and interactions with strangers.

IDEOLOGIES AND ECOLOGIES OF BELIEF

Senses of the Term *Ideology* and Their Relation to Ecologies of Belief

There are many senses of *ideology*. First, there is De Tracy's original sense in which ideology is the rational investigation of the origin and basis of ideas.[2] Second, there is Marx's most prevalent sense in which an ideology is a system of ideas that serves to perpetuate power relations.[3] Third, there is Lenin's (and, to a significant extent, Lukacs's) sense in which an ideology is the expression of a class's social position and desires.[4] Fourth, there is Martin Selinger's sense in which an ideology is any and every action-oriented set of beliefs organized into a coherent system.[5] Fifth, there is Clifford Geertz's sense in which an ideology is a cultural system.[6]

In which of these or any other sense are ideologies similar or identical to ecologies of belief? First, ecologies of belief are embedded in larger—social— ecologies but are not identical with entire cultural systems, which may involve a number of mutually interacting social ecologies. Hence, ecologies of belief are not ideologies in Geertz's sense. Second, given that ecologies of belief are not necessarily coherent but often involve tensions between incompatible beliefs and, further, may, but need not, be action-oriented, they are not ideologies in Selinger's sense. Third, given that ecologies of belief may make room for meaningful dialogue and critical scrutiny, they are not ideologies in Lenin's or Lukacs's senses. Fourth, given that ecologies of belief are interactive networks of beliefs that make room for, but are not identical with, inquiry about some or other of these beliefs, they are not ideologies in De Tracy's sense. For, in this latter sense, ideologies are forms of investigation. Fifth, since ecologies of belief may, but need not, be linked to control or domination, they are not ideologies in Marx's sense.

To summarize, ecologies of belief are narrower than entire cultures; may, but need not, be action-oriented; may be, but often are not, coherent; and, though not identical with critical scrutiny and inquiry, they often make room for it. Of course, in some additional sense, some ideologies could make room for critical scrutiny and inquiry. But why expand the sense of *ideology* in such a manner?

One may adopt such expansion, so long as two things are kept in mind. First, not just any, nor indeed any frequently recognized, ideology is tolerant of meaningful dialogue and critical scrutiny. For example, the ideologies involved in the religious fundamentalism of Christian or Islamic sects that adopt a literal interpretation of the Bible or the Koran, and are supposed to remain unchanged, are thus not tolerant.[7] Second, the said expansion should not eliminate the focus on power imbalances, which is where the term *ideology* seems most helpful. Hence, in accordance with this essay's purposes (indicated in the questions raised at the outset), I will focus on these features and characterize ideology accordingly. This leads to the essay's first thesis: For the present inquiry's purposes, ideology is most usefully understood as a network of beliefs, attitudes, conventions, practices, and institutional priorities associated with different, sometimes mutually conflictive, distributions of power and resources.

Given this, it is possible that some, though not the most frequently discussed, ideologies make room for and encourage meaningful dialogue and critical scrutiny. Hence our second thesis: Ideologies may make room for and encourage meaningful dialogue and fruitful interaction in the process of social change. That is, some ideologies may be *benign* in that they are open to reason and any changes it may motivate.

Even in the expanded sense just indicated, though ideologies overlap with ecologies of belief and are embedded in some of them, they do not exhaust them. There are at least two reasons for this. First, ecologies of belief range over not merely uneven and conflictive, but also even and nonconflictive distributions of power and resources, while ideologies, in the sense used in this essay, focus only on uneven ones. This reason also serves to distinguish ideologies from social ecologies, because the latter, but not ideologies, range over not merely uneven and conflictive, but also even and nonconflictive, distributions of power and resources. Second, ecologies of belief need not be, while ideologies always are, action oriented. Hence, this essay's third thesis: Ideologies are not identical or equivalent to ecologies of belief or, for that matter, to social ecologies.

A crucial cluster of questions arises here: How should we go about engaging in research or other activities in a manner that, if ideological, falls within the scope of a benign ideology? What ways of bringing about changes in beliefs, and their associated attitudes and social structures, are justified?

Will the identification of these ways help establish what approaches, attitudes, practices, policies, and social structures, scientific or not, gain legitimacy from their making room for, and perhaps also encouraging, legitimate ways of bringing about changes in beliefs, attitudes, practices, policies, and social structures? I will next address these questions.

Ideology and Legitimacy

Jürgen Habermas has significantly devoted himself to a discussion of ideology and legitimacy. For him, the study of ideology is the study of systematically distorted communication, and the criterion by which to judge an ideology is the extent to which it makes room for undistorted communication or reasonable discourse. In other words, an ideology is legitimate to the extent that it makes room for a "cooperative search for truth in which only the force of the better argument prevails."[8]

Given this criterion, Habermas suggests a critical question to ask about any social system and the ideology it reflects:

> How would the members of a social system, at any given stage of development of productive forces, have collectively and bindingly interpreted their needs (and which norms would they have accepted as justified) if they could and would have decided on the organization of social intercourse through discursive will-formation, with adequate knowledge of the limiting conditions and functional imperatives of their society?[9]

Habermas makes his position quite clear:

> We cannot explain the validity claim of norms without recourse to rationally motivated agreement or at least to the conviction that consensus on a recommended norm could be brought about with reasons.[10]

This position has prompted a variety of criticism. An important one is that equating ideology with distorted communication is not specific enough because not all distorted communication is ideological. Habermas could reply that all *systematically* distorted communication is ideological. But this is not clear. As discussed in essay 9, there seems to have been systematically distorted communication throughout much of Faraday's public lectures. Yet, it was not ideological in Habermas's sense—or in our sense—of *ideology*, but simply aimed at avoiding unnecessary controversy about views Faraday had not yet fully worked out.[11]

This points to a more crucial objection: The whole concentration on language and communicative competence may be suitable for discussing changes in scientific beliefs. Yet, when dealing with changes in overall social ecologies, it tends to neglect inequitable distributions of power and resources: Access to undistorted communication may not be enough to secure equitable social arrangements. After all, consider the Mafia. When it makes an offer we cannot refuse, communication is crystal clear, entirely undistorted. Power distribution, however, is hardly equitable, and the resulting social situation hardly has any legitimacy.[12]

A different type of objection concerns the role of argument and reason in Habermas's conception. For example, Jon Elster, who is largely in sympathy with Habermas's views, fears "that it might be dismissed as Utopian, both in the sense of ignoring the problem of getting from here to there, and in the sense of ignoring some elementary facts of human psychology."[13] Accordingly, his objections focus largely on how difficult it is to reach a consensus and how questionable it is to seek it or even to approach it, let alone to require that everyone be involved in the task. Elster has a more cautious view of the place of consensus in working out social arrangements.

> It is the concern with substantive decisions that lends the urgency to political debates. The ever-present constraint of time creates a need for focus and concentration. . . . Yet within these constraints arguments form the core of the political process.[14]

This sensitivity to time constraints is a move in the right direction, but, as I have extensively argued elsewhere, it still is a somewhat simplistic doctrine. The constraints on arguments in politics go far beyond matters of time. Some policy and decision problems are intractable because of the type of social conflict that poses the problem—say, a confrontation—or because of other social circumstances—the existence of issue-overload. Even if time were not of the essence (which it often is), seeking consensus might be an irrelevant way of dealing with the problem. It might amount to giving reasons for some belief, decision, or social arrangement to those who are set on not listening to reason at all.[15] I will next outline a more realistic approach.

NATURE, SCOPE, AND LIMITS OF THE PHILOSOPHY AS DIPLOMACY APPROACH

In actual social life, simplistic approaches such as that proposed by Habermas or the modest modification suggested by Elster often amount to recipes for catastrophe. Instead of such approaches, we should use criteria for evaluating a range of responses to given types of situations. The criteria should specify conditions under which beliefs may be upheld, decisions may be made, conventions may be settled, practices may be introduced, policies may be instituted, and social arrangements may be established *given existing time constraints; polit-*

ical pressures; economic, technological, and other relevant conditions; the people involved; the information they can be expected to have at the time; and the moral and political implications of upholding the beliefs, making the decisions, settling the conventions, introducing the practices, instituting the policies, or establishing the social arrangements.

In *Philosophy as Diplomacy*, I have formulated such criteria for policy and decision making.[16] These criteria also apply to other—but not all—of the responses just listed. First, the criteria also apply to the introduction of practices, because they apply to decisions, and the introduction of a practice is characterized by a great number of decisions generalized throughout a social group. Second, the criteria also apply to the establishment of new institutions and social arrangements, for they are characteristically established through a combination of policy making, decision making, and practice introduction, and the criteria apply to each of these components.

So far as they apply to decisions, the criteria apply to the decision to uphold given beliefs publicly. However, they do not serve to justify the beliefs themselves. That is, the criteria I will outline apply to decisions that concern, for example, the stand we take about predominant social ecologies—say, those affecting the institutionalization and organization of science in society. However, they do not serve to establish what to consider plausible hypotheses for research or what to accept as confirmed in the process of scientific inquiry. These latter matters of belief may be upheld by particular researchers or groups of researchers moved by highly subjective, when not arbitrary, motives—from their idiosyncratic faiths to a desire for prestige—but, in the ongoing process of inquiry, they are primarily resolved on the basis of such criteria internal to the inquiry as clarity, coherence, completeness, informative power, explanatory power, and predictability.

Also, new attitudes may develop—and, indeed, become necessary—in the policy- and decision-making process, and the criteria apply to the decisions involved in developing the attitudes. Yet the criteria do not apply to the attitudes themselves. Given these constraints, I will next formulate a list of questions that can help guide our evaluation of envisioned responses to the social and personal situations we may face. These concern social ecologies, our personal development in their midst, and our views concerning the ideologies and whole ecologies of belief involved in those situations.

UNDERSTANDING THE SITUATIONS WE FACE[17]

When faced with the entire gamut of responses subject to evaluation—from decisions, through attitudes and beliefs, to practices, policies, institutions, and social arrangements—questions such as the following are pertinent:

1. Is the situation in which the item is to be evaluated predominantly conflictual or consensual? That is, does the situation, as in the case of abortion, involve sharp differences of opinion or conflicts of demands? Or, as in the case of determining the number of commemorative stamps to be issued in a given year, is the situation largely free from conflict?

2. Does the situation predominantly make room for appeals to reason and meaningful dialogue—that is, is it predominantly rational, however loaded with controversy—or, as in the case of wars, is it mainly confrontational?

3. Does the situation arise at the macrolevel or microlevel of social life? Does the situation involve a local problem—such as that of deciding how much money to allocate for education in a town's budget? Does the situation involve a regional problem—such as how to ensure that New England's energy needs are met? Does it involve a national problem— such as what to do about Medicare in the United States in the mid-1990s? Does the situation involve an international problem—such as what to do about acid rain that significantly originates in the United States Midwest but partly falls on Canada? Does the situation involve a global problem—such as what to do about increasing levels of atmospheric carbon dioxide and the greenhouse effect likely to result?

4. Is the situation destabilizing one or more groups—or possibly entire societies, as in the 1990s war in the former Yugoslavia—or not?

5. Does the situation involve violations of individual rights, and, if so, what is the nature and extent of these violations?

There are, no doubt, additional questions worth asking. Yet the ones just formulated are invariably crucial to achieving a politically and morally sound perspective on the practical problems posed by the situations faced. All of them ought to be asked, though not necessarily in the order in which I have formulated them. However, the order I chose reflects a manner in which the problems can be more easily understood in all their complexity, and the approach I am here outlining can be more directly put to work.

People, however well-meaning, often get caught in situations in which they argue merely about who has the right to what or, alternatively, what will or will not work. In this manner, they lose perspective. Whenever this happens, shifting attention to such things as whether the situation is conflictive or not; whether it is local, regional, national, international, or global; and whether it makes room for appeals to reason and meaningful dialogue, leads to a better understanding of the problem. This in turn helps assess the relative weight of the problem's features from a social and personal standpoint— including whether individual rights or societal stability take precedence in dealing with the problem.

No doubt, the approach I have just outlined does not guarantee certainty. Nor will it work in a society (fortunately only an imaginary society) where most people are idiots, lunatics, crooks, opportunists, or utterly callous. However, it has a good chance of helping ordinary people understand the various types of situations that arise in societies as we know them. Of course, understanding is only the beginning. The question still remains: How to respond? I will next begin to list the questions to ask in this regard.

EVALUATING ENVISIONED RESPONSES

Once one has achieved a sense of the situation's overall complexity—for example, for the overall complexity of disposing of chemical or nuclear wastes—the following additional questions are crucial: First, what responses are likely to be both feasible and effective in dealing with the situation? Second, which of these responses—for example, what waste disposal site selections—are equally or more likely than alternative ones to lead to undesirable consequences to groups or to limitation on individual freedoms and well-being?

Having identified the responses that could have such undesirable implications, however, is not automatically sufficient for ruling them out. For those affected or, if these are not in a position to judge, their trustees or representatives, may not find them intolerable. After all, some towns do not find burying other towns' PCBs in their own backyard in exchange for cash intolerable. And, if they do not, though we can think their judgment mistaken or misguided, who are we to treat their judgment as anything other than crucial in the process? Hence, we need to ask at least a third question: Are the individuals likely to be affected, or are their trustees or representatives, when of sound and cool mind, free from the influence of coercion and manipulation, and well informed, likely to find the said social consequences or limitations of their freedoms and well-being intolerable?

If the answer to this question is negative, then regardless of how unpalatable to our idealism the responses may be, they are nonetheless permissible. We may, at the same time, try, through dialogue, to change the minds of those who find it tolerable, But the point remains that, in and of itself, our point of view takes no precedence over theirs. We are in no way automatically entitled to treat them as needing liberation from their arguably unenlightened ways despite themselves, let alone to attempt to liberate them from these against their will, or to liberate some of them at the cost of sacrificing others, for the sake of the cause of utter liberation from ignorance or any other cause.

If the answer, however, is affirmative, and those affected or their trustees or representatives find the envisioned responses intolerable—and, in the environmental examples previously mentioned, this not-in-my-backyard response is most common—then the said responses are not automatically permissible either and a new situation, posing a second order problem develops. It can be formulated as follows: How can we go about settling the sharp differences that divide those who confront the previous situation concerning tolerable solutions to it? This is a second order problem because it calls for social decision procedures that focus on the division and differences, not the initial subject matter at issue which created the division and differences in the previous situation.

TYPES OF SOCIAL DECISION PROCEDURES

In the situation just described, one must ask, What policy-making procedures are likely to be both feasible and effective, given the existing time and budgetary constraints, degree of uncertainty about the procedures' consequences, and probability of their helping develop public support for some of the still-contested responses or at least develop acquiescence to them? How much time, money, and energy is there? How much should be devoted to dealing with the sharp divisions of the said envisioned responses? If the conflict is a confrontation, through what social decision procedure or combination of procedures is it permissible to address it? Which ones are permissible if it is a controversy?

In dealing with confrontations and their characteristic breakdown, a variety of procedures suggest themselves: outflanking, strikes and demonstrations, threats, war. None of these procedures relies on meaningful dialogue and appeals to reason, though each may be applied by using reasons. Yet the procedures may be politically sound. And they may be morally permissible if the conflict is a confrontation and they are accompanied by requests or demands to reestablish the dialogue. Indeed, the less reliant on reason and meaningful dialogue the situation posing the problem is, the more these procedures are permissible—at least so long as they are accompanied by efforts to bring reliance on reason and meaningful dialogue back into the picture.

For example, Iraq's invasion and annexation of Kuwait in August 1990 was met with nearly unanimously supported UN sanctions while negotiations were still being sought.[18] Whatever the weaknesses of the particular sanctions, there was little or no disagreement throughout the world community that they were politically and morally unobjectionable, so long as they were applied in the search for peace, not for strategic advantage or political or economic gain.[19]

If the conflict is a controversy, however, none of these procedures appears permissible, if at all relevant. Instead, such things as political insulation, arbitration, voting, negotiation, mediation, consensus building, and discussion of merits—which make room for reliance on reason and meaningful dialogue—take center stage. And these procedures are different from those called for by societally disruptive situations in which nonrational or mistaken consensus is predominant. In that case, procedures such as demonstrations, sit-ins, boycotts, and other consciousness-raising activities appear most suitable.

WHAT PROCEDURES TO USE WHEN?

Given the preceding discussion, the variety of procedures just listed may be used in the following order of priorities: First, just as no one is entitled to exclude the competent judgment of those affected from consideration, no one is entitled to engage in confrontation when nonconfrontational procedures are feasible and likely to be effective. Only the latter are permissible under the circumstances. The use of terrorist means to call attention to one's ends might be feasible and effective. But it would not be permissible under the circumstances because it would move us away from, not, as should be the case, closer to reliance on reason and meaningful dialogue.

Second, if a confrontational situation develops, only nonviolent confrontational procedures, if likely to be feasible and effective, may be used. For violence breeds violence, and, as in the terrorist's previous case, it carries us away from, not, as should be the case, closer to reliance on reason and meaningful dialogue.

Third, if only violent procedures are feasible and likely to be effective, then only the least violent of these may be employed and no more than to the extent they are necessary (an extreme case under this category is that in which only combat and war are feasible and likely to be effective) to make the situation nonconfrontational. That is, terrorism for terrorism's sake, just as war among established states for the sake of war, is unjustified. The reasons for this are the same as those formulated in the previous two cases. If the use of violence is ever justified, it must be both feasible and likely to be effective in bringing us closer to a situation in which reliance on reason and meaningful dialogue take center stage.

This is not to say that violence is justified whenever it is likely to restore order. For order may be oppressive, hence, not a form of peace but a quiet form of war. If violence—whether that of nation-states or that of terrorists—is ever justified, it must at least be likely to bring us closer to nonconfrontational situations.

Of course, there are situations in which nothing works. This leads to a fourth condition: If none of the preceding procedures is feasible or likely to be effective, then coping and waiting (as the United States did for many years concerning Cuba), perhaps accompanied by discussion of merits, consensus building, and other feasible procedures (such as an economic embargo) that may be marginally effective in the long run, may be used until circumstances change.

At any rate, transformations in a given group's social ecology at a given time can be brought about in accordance with the previous conditions. These, as indicated, extend the pragmatic approach developed in my *Philosophy as Diplomacy* beyond policy and decision making to a variety of responses ranging from practice introduction and convention settling to the establishment of new social arrangements. That is, and this is the essay's fourth thesis, the transformation of the social ecologies and the ecologies of belief embedded in them should proceed along the pragmatic and open-ended lines of a philosophy as diplomacy.

This is a more promising approach regarding not merely social ecologies but also our views concerning ideologies and ecologies of belief. For it gives intellectual activities such as scientific and philosophical inquiry enough leeway so that the scientific and philosophical beliefs they address can be evaluated and, if need be, changed, in accordance with the inquiries' protocols. This provides grounds for inferring the essay's fifth thesis: The philosophy as diplomacy approach is suitable for soundly dealing with people's views concerning changes in ecologies of belief. It commits science and philosophy neither to the intellectual suicide of adopting a creed, as feared by T. H. Huxley, nor to the practical suicide of being cut off from the rest of human life. For it makes room for social and individual intellectual life between the excesses of social order, which always limit individual freedoms, if not well-being, and the excesses of social disorder, which always limit well-being, if not individual freedoms. Freedom of thought and inquiry, as well as human well-being, stand to gain in the process.

NOTES

1. Robert Andrews, *The Concise Columbia Dictionary of Quotations* (New York: Columbia University Press, 1989), p. 233.

2. Antoine Destutt De Tracy, *Elements d'Ideologie*, discussed in David McLellan, *Ideology* (Minneapolis: University of Minnesota Press, 1986), p. 6.

3. Karl Marx, *Selected Writings* (Oxford: Oxford University Press, 1977), p. 389f., 172. See also McLellan, op. cit., pp. 14–18.

4. V. Lenin, "What Is to Be Done?" *Selected Works* (Moscow: Progress Publishers, 1963) 1: 156f., and G. Lukacs, *History and Class Consciousness* (London: Merlin Press, 1971), p. 258f. See also McLellan. op. cit., pp. 24–25.

5. Martin Selinger, *Ideology and Politics* (New York: Free Press, 1976), passim. See also McLellan, op. cit., p. 82.

6. Clifford Geertz, *Ideology and Discontent* (New York: D. Apter, 1964), passim. See also McLellan, op. cit., pp. 82–83.

7. For a discussion of this and related types of ideologies, see Mario Bunge. *Seudociencia e ideología* (Madrid: Alianza Editorial, 1989), pp. 125–33.

8. Jürgen Habermas, "Morality and the Law," in Sterling McMurrin, ed., *The Tanner Lectures on Human Values, VIII* (Salt Lake City, Utah, and Cambridge: University of Utah Press and Cambridge University Press, 1988), p. 243.

9. Jürgen Habermas, *Legitimation Crisis* (Boston: Beacon Press, 1975), p. 113.

10. Ibid., p. 105.

11. For an objection along these lines, see McLellan, op. cit., p. 79.

12. This type of objection is also indicated by McLellan, loc. cit.

13. Jon Elster, "The Market and the Forum: Three Varieties of Political Theory," in Jon Elster and Aanund Hylland, eds., *Foundations of Social Choice Theory* (Cambridge: Cambridge University Press, 1986), p. 114.

14. Ibid., pp. 114–20.

15. See my *Philosophy as Diplomacy* (Amherst, N.Y.: Humanity Books, 1994), especially "Issues and Issue-Overload: A Challenge to Moral Philosophy," pp. 1–14; "A Delicate Balance: Reason, Social Interaction, Disruption, and Scope in Ethics and Policy-Making," pp. 87–111; and "Practical Equity: Dealing with the Varieties of Policy and Decision Problems," pp. 112–50.

16. Ibid., passim.

17. For a more thorough discussion of the points in the essay's remainder, as they concern policy and decision making, see my *Philosophy as Diplomacy*, pp. 105–106, 112–35.

18. See Combined Wire Services, "UN Imposes Embargo on Iraq," *Hartford Courant*, August 7, 1990, pp. A1, A4.

19. See Ellen Warren, "Bush, Gorbachev United Against Iraq," *Hartford Courant*, August 7, 1990, pp. A1, A4.

Dialogue

— I am skeptical about your account's implication that the only difference between such ecologies of belief as science and philosophy, on the one hand, and other ecologies of belief is simply a matter of attitudes.

— Your description of the situation puzzles me. There are, no doubt, scientific, as well as philosophical, ecologies in the sense in which I have been using *ecologies of belief*. They belong in a subclass of ecologies of belief I called *intellectual ecologies*. But they are not entirely cut off from ordinary beliefs. Rather, they are part and parcel of the overall ecologies of belief embedded in the social ecologies predominant in given societies.

— I see; but my skepticism remains. Why should intellectual ecologies differ only in attitudes from other ecologies—perhaps from religious ecologies—also embedded in the overall ecologies of belief?

— I never said they do, though, I grant you, intellectual ecologies tend to be associated with attitudes that are not prevalent in nonintellectual ecologies.

— If not simply attitudes, what is the difference?

— The difference, as I indicated before, is also one of predominant protocols. Religious beliefs are not predominantly or primarily, if at all, established by appeal to such criteria as clarity, coherence, completeness, informative power, explanatory power, or predictive power. By contrast with such beliefs, these criteria are central for establishing intellectual beliefs.

— OK. But what good are these criteria without a critical and self-critical attitude on the part of those who use them?

— No good. You're right. But this is only to say that such an attitude is crucial in the practice of science and philosophy. It is in no way to say that such an attitude is all that distinguishes intellectual activities and their products, intellectual ecologies, from other sorts of ecologies of belief.

— Fine. But this raises another concern: Is this critical and self-critical attitude characteristic of scientific and philosophical inquiry a ground for establishing beliefs?

— Having such attitudes, in and of itself, establishes nothing. It only sets limits to the procedures by which one can go about establishing beliefs.

— Couldn't the fact that a critical and self-critical attitude is central to scientific and philosophical inquiry, however, serve to establish, for example, the belief that a procedure whereby we can accept any belief we desire to hold is inadequate?

— Yes, but you are falling prey to an ambiguity. In the case you describe, what serves as a ground for establishing the said belief is not the attitudes themselves but the fact that these attitudes are central to scientific and philosophical inquiry.

— I see your point; but let me try a slightly different approach. Can't the said critical and self-critical attitude, in setting limits for how to establish beliefs, be a ground for believing that certain—say, desire-based—beliefs are not established?

— I don't think so. They may lead us to inquire what grounds there are for accepting such beliefs. But the belief that, at a certain point in the inquiry, there are no such grounds will be based on the fact that none are then available except for the desire to hold such beliefs and that desires are not sufficient for establishing the truth of beliefs.

— Can't desires, however, sometimes be grounds for holding such beliefs? I think that William James had a point when, in "The Will to Believe," he held that, in certain extreme situations—whenever one faces a genuine option, evidence or proof are mixed or absent, and one's happiness is contingent on one's belief—one has the right to believe as one pleases.

— I agree. But, in fact, the very possibility of establishing James's belief in this thesis is grounded on the very possibility that reasons other than mere desires can be given about it and that critical and self-critical thinkers— among them William James—have offered them.

— We seem to be slowly sliding into a discussion of moral matters.

— We certainly are. I intend to address them in the next essay.

— Good. I have been wondering how you can deal with the relations between attitudes and beliefs we just discussed when it comes to morals and mores.

— Then you should be interested in the next essay. Let us proceed.

A Balancing Act: Morality, Moral Ecologies, and Moral Philosophy

BETWEEN AGGRESSION AND CIVILITY

Humans are among the most irritable, aggressive, and individualistic creatures on earth. Yet, they live in societies where they generally follow customs—like that of eating (or, alternatively, shunning the eating of) insects —laws—from those regulating motor vehicles, through those establishing property rights, to those punishing violent crimes—and a variety of moral injunctions that go well beyond custom and the law and enjoin us to do such things as showing gratitude and giving to charity. How are these customs, laws, and moral injunctions related to each other? Where do they fit in relation to morals, mores, morality, moral philosophies and theories, and moral philosophy? What role do they have in the existence of groups and the lives of individuals? Are they part of moral ecologies? In what sense of *ecologies*, and how? Can we understand, evaluate, and apply moral philosophies and theories along ecological lines? How?

These are the questions I will address in this essay. I will advance the following theses: First, customs can exist where there are no laws or morality. Second, a group's customs and morality partly make up its mores. Third, laws may be argued to introduce a morally justified component in a social group; but they cannot and should not completely take the place of morality. Fourth, moral philosophies and theories are best understood, evaluated, and applied by reference to *moral ecologies*—to networks of such things as morals, mores, laws, interpersonal influences, political pressures, and other components of *social ecologies* that the people involved consider crucial to individuals' lives— and are open to critical scrutiny and interactions in the social testing process.

In advancing these theses, I will further clarify the notion of social ecologies discussed in previous essays, discuss how it relates to moral ecologies, and characterize the social testing process in some detail. Let us start by clarifying the moral notions that are the focus of these theses.

184

From Custom to Moral Philosophy

Morals at Work

> A twenty-nine-year-old married, professional woman finds out she is preg-
> nant again four months following the birth of her third child. She asks her
> physician whether there are any health reasons why it is not advisable for
> her to have this child, and the physician tells her there are none. She dis-
> cusses her pregnancy with some close friends and relatives, all of whom tell
> her that they love her and will support her whether she decides to have the
> child or not. She believes an abortion is taking a potential human life and
> that, except when there are clearly overriding moral reasons, abortion is
> wrong. Yet, the recent birth of her third child has created some financial
> strains in her family. She must work full days to make ends meet, and, after
> putting in an eight-hour day, she hardly has the time and energy to take
> care of the newborn, plus the two older children. In addition, she does not
> at all feel ready to have still another baby. She thinks: "An abortion is
> taking a life; but I also ought to think of the lives of my four-month-old,
> and my older children, and of my own life. And these would be short-
> changed, if not harmed, were I now to have another child."

Universal agreement on whether she would be justified in having an
abortion is unlikely. Some people say that abortion is killing an innocent
human being, hence always wrong. Others hold that it is sometimes right,
though they differ on the conditions under which it is right. Still others deny
that abortion is murder at all and hold that it is always morally permissible.
Conflicting statements like these are not unusual. One individual often holds
beliefs about right and wrong, good and bad, justified and unjustified, that
conflict with those of another individual. That is, personal ethics, a person's
morals, vary from person to person and often, with the passage of time,
change in the same person. The fact that a person's morals vary does not settle
the question of what action is right, but it does provide an example of one of
the ways in which the terms *ethics* and *morals* are used: the *personal* sense, in
which someone's ethics or someone's morals are a particular person's beliefs
and presuppositions about right and wrong, good and bad, justified and unjustified. This—a person's morals—are something a particular person *has*.

Scope of a Person's Morals

As this characterization makes plain, a person's morals include, but are not
exhausted by, the person's *beliefs* about right and wrong, good and bad, justified and unjustified or, in short, the person's moral beliefs. A person's morals
also include the person's moral *presuppositions*. What does this term cover? First,

moral presuppositions include what is taken for granted by such things as *state-ments and questions*—for example, that you have a spouse and have been abusing your spouse, when one asks you: Have you stopped abusing your spouse?

Second, moral presuppositions also include what is taken for granted by *actions*. An example is the presupposition that you have been abusing your spouse and this behavior is blameworthy, when I blame you for it. Third, moral presuppositions cover what is taken for granted by *traits*. An example is the presupposition that increasing one's awareness of others' needs is morally desirable when, in displaying such traits as kindness and fairness, a person tries to become more sensitive to the needs of others.

That is, a person's morals include the person's moral beliefs, as well as a variety of things the person takes for granted in everyday life—from morally significant facts to moral rules, laws, principles, values, or ideals displayed in the person's behavior or involved in the person's character.

Morals and Immorality

The preceding discussion might be taken to suggest that people's actions, motives, attitudes, and traits are always in harmony with their morals. But they need not be, and sometimes are not. For example, no such harmony is present whenever one acts contrary to one's moral beliefs—say, when someone breaks a promise while believing there is no excuse for it and, as a result, feels ashamed or guilty. Nor is there any such harmony whenever one has feelings, or wishes, contrary to one's moral beliefs and characteristic atti-tudes—say, when carried away by anger, one wishes, or even sets out to do, serious harm to someone one knows to be undeserving of, and one is not typ-ically disposed to make the victim of, any such harm.[1]

Indeed, such lack of harmony between one's morals, on the one hand, and such things as one's conduct and one's motives, on the other, need not be out of character. For a person whose morals enjoin kindness and fairness might, as a matter of character, be so irritable and temperamental as to have unkind and unfair motives, and even act unkindly and unfairly quite often, and then feel regret, shame, or guilt about it.[2]

Mores

In the previously described abortion example, a person's morals may include the belief that, under the conditions described, it would be morally permis-sible for the woman to have an abortion. They may also include, perhaps as grounds for this belief, the presupposition that some considerations are so per-sonal and private as to be best judged by those directly involved, in particular, by the woman herself. Accordingly, one would think that the woman would be the best judge of whether, under the circumstances, the lives and well-being of her children, as well as her own, take precedence over the life of the fetus. This is not an unusual view, and it, or something very much like it, is reflected by the United States *Roe* v. *Wade* and related abortion decisions.[3]

Of course, the fact that an opinion is reflected in the law does not make it right. What the law says in no way settles the question of what it ought to say. Yet, what the law says is not irrelevant to the moral description of both the woman's problem and the conduct in addressing it. Nor is it irrelevant for evaluating this conduct. For what the law says—what it permits and forbids—is part of both the circumstances that contribute to create a problem for the woman and those in which her conduct in fact must address the problem. With this in mind, let us return to our case.

> Aware of the conflicting opinions different people hold about abortion and, partly for this reason, feeling uneasy about it, the woman proceeds to take a look at the codes of conduct of various religious and professional groups involved in abortion discussions. Some religious codes strictly prohibit all, or nearly all, abortions. Others are somewhat permissive. The professional codes she finds tend to be somewhat noncommittal. The woman soon realizes that, whatever guidance a certain group's code of conduct may provide, it is, like the law's guidance, criticized by disagreeing members of the group or some of the general public.

A group's code of conduct cannot, by itself, settle the question of what actions are right. However, such a code constitutes a partial example of ethics in the social or group sense of the term *ethics*. In this sense, ethics goes beyond conduct, also regulating such things as attitudes and practices, and, in this comprehensive sense, it is a group's *mores*: a particular group's predominant beliefs and presuppositions about right and wrong, good and bad, justified and unjustified. Accordingly, ethics in the social sense is something a group—rather than a particular individual—*has*, in that the group has explicitly, however partially, formulated it or predominantly holds it.

Mores, Manners, Custom, and Morality

A group's mores may not distinguish between manners (for example, thanking those who let us go out of an elevator first), matters other than manners that are not morally significant (for example, having pasta on Sundays among Italian descendants in Argentina), and matters of moral significance (for example, acting contrary to the rules against murder). Simplistic mores such as those here envisioned acknowledge only one category: *the done thing*. This leads to our first thesis: Customs can exist where there are no laws or morality.

However, mores need not be so simplistic. They may distinguish between various types of customs: those that concern mere manners; those that concern matters other than manners yet are not morally significant; and those that are morally significant. However, even when such complex distinctions are present, mores may be quite parochial and uncritical. For example, in ancient Greece, burying the dead was a morally significant custom; cremating them was a morally significant practice in Persia; and ancient Greeks and Persians shocked each other when they observed each other's practices concerning the dead.[4]

Yet, when groups with conflicting mores come into contact, more critical attitudes and judgments often develop among members of each group. For when the beliefs and presuppositions constituting a person's or group's ethics are in conflict with those of others, giving rise to disagreements between individuals, groups, or between groups and some of their members, the question of who is right characteristically arises. And sometimes, upon facing up to the said conflict, people become intellectually cautious and self-critical. This is where mores, in effect, make room not just for *a* morality—that is, the morality of a particular group—but for the general concept of morality—the standpoint from which any morality can be critically scrutinized. For critical questions can be, and actually are, asked about some of the morally significant matters the mores involve. Whether they are asked or not, our second thesis still holds: A group's customs and morality partly make up its mores.

Law, Laws, and Morality

One might be inclined to think that a society makes room for morality in at least one other way: by introducing a legal system. By itself, however, such a system will not do. To be sure, as H. L. A. Hart made plain, in evolving from a situation involving only custom, a legal system would simply systematize the manner in which morally significant aspects of a social group's mores are to be addressed through official lawmaking, judging, and enforcing bodies.[5] To the extent that it is effective, such a system introduces some *desirable order* in society, by doing away with such things as overbearing traditionalism and spasmodic mob rule. But it does not follow that a legal system will make room for morality beyond such minimal changes. For example, a legal system may—and some have—become stagnant, unable to make adaptive corrections sensitive to social—say, demographic or technological—change, which often leads to generalized enforcement failures and loss of legitimacy.

However, even if any legal system—the law—makes significant room for morality beyond the bare minimum of doing away with stagnant traditionalism and mob rule, it cannot entirely codify morality. Let us see why.

First, there are significant differences between legal and moral entitlements, not the least of which is that legal entitlements depend on rules of legal evidence, while moral entitlements need not do so. Consider the following case:

> Mrs. Mary Morand had been issued a certificate of deposit for $14,000 to the order of herself or Mr. John Gutre. She had told third parties that she meant to give Gutre the proceeds as a gift, but she kept control of the document until her death. Morally, there is good reason to believe that, upon Mrs. Morand's death, Mr. Gutre deserved to receive the gift she had promised. Legally, however, things work otherwise. Gutre sued for the proceeds and lost, on the grounds that, since Mrs. Morand had never given up control of the document, the gift had never taken place.

Though morally one may be entitled, legally one is not entitled to gifts announced but never made. In order to rule otherwise in the preceding case, hearsay would have to be admitted as legal evidence—a sure recipe for legal chaos.[6]

Second, laws could not effectively cover all morally significant areas. Imagine the enforcement nightmare that would be created by laws meant to regulate ungrateful, unkind, or excessively generous behavior! Given the court overload caused by laws that fall far short of regulating such matters, the said additional laws would lead to legal paralysis, hence undermine the legal system.

Third, even if such laws could be enforced, they should not. For enforcing them would be a paradigm of morally meddlesome and busybody behavior. This would not be simply an invasion of privacy and unjustified limitation of individual freedom. Through its meddlesomeness, it would also tend to undermine the legal system. Hence, on both counts, it would be morally objectionable.

Fourth, the morality of any laws themselves can be questioned. As Kurt Baier put it:

> When we ask whether something is really in accordance with or contrary to a given society's custom or law, this is just another formulation of the . . . question: what is *rightly* thought to be in according, what contrary to a society's custom, law, or morality? But in the domain of morality, we can ask a question which has no analogue in the domains of custom and law: "Is what is (rightly) correctly thought to be according/contrary to a given society's morality, rightly (correctly) thought to be according/contrary to morality?"[7]

The mere fact that this question is meaningful indicates that the standpoint of morality is not reducible to mere custom, law, mores, or, more specifically, any particular group's morality. This point is further strengthened by the fact that reasons can be, and often are, convincingly given for or against alternative answers to the preceding question. Hence, our third thesis: Laws may be argued to introduce a morally justified component in a social group; but they cannot and should not completely take the place of morality.[8]

Moral Philosophy

As soon as a question such as that just discussed is raised, moral inquiry begins. This is not a typically detached activity. Certainly, in cases like the previously discussed concerning abortion, the disagreements are often sharp, issuing in controversies and even confrontation, and creating an urgency in establishing who is right as well as in acting on the problem of resolving such sharp disagreement. When people engage in critical inquiry about such matters of disagreement they *do* ethics or moral philosophy rather than simply have ethics or morals or a moral philosophy that may (or may not) be shared with most members of their group thus being (or not being) in harmony with the group's mores.

In this latter sense, in which people do rather than simply have a moral philosophy, ethics or moral philosophy is an activity, not simply a set of beliefs and presuppositions. The activity is not identical with but *about* beliefs and presuppositions, and it is often prompted by conflicts among them. Thus ethics becomes a branch of inquiry—sometimes also called ethical theory, moral theory, or reflection on morality. It is a critical study with the goal of soundly dealing with problems of right and wrong, good and bad, justified and unjustified, that arise in people's lives.[9]

KINDS OF MORAL PROBLEMS AND ISSUES

The abortion case previously described raises the question of whether it would be right for the woman to have an abortion. This is a question about the rightness or wrongness of a particular individual's action under specified circumstances. It poses a moral problem of *conduct*, for it is about an action or piece of conduct.

In discussing the rightness or wrongness of the woman's having an abortion, the question may arise whether actions of this type should be permitted by law. The focus of this question is not an action but a social policy. The problem it poses is not of conduct but *institutional*. It concerns the justifiability of such things as policies (for example, those about abortion), practices (for example, that of using only masculine pronouns to refer to persons), or institutions (for example, the institutions of marriage, business, art, and the state).

There is yet a third kind of ethical problem that can be raised in discussing the abortion case. The woman who considers the option of having an abortion may ask, What is having an abortion going to do to me? What kind of character traits would such an action tend to instill in me? Are they good traits? The focus of these questions is not a policy, practice, or institution, nor is it primarily an action. It is, rather, character traits specific to a person, such as courage and kindness—which are to a person's credit—and cowardice and callousness—which are to a person's discredit—and which, together, make up a person's character. Such questions about character traits raise ethical problems of *character*: problems about the goodness or badness of a person's character, character traits, motives, and attitudes.

Not all kinds of moral problems, certainly not all kinds of institutional moral problems, prompt the same responses. Some, like the problem of how many and which persons to honor by means of commemorative stamps in a given year, involve little, if any, controversy. By contrast, other problems— say, whether there should be public funding for abortions—involve a great deal of controversy, when not outright confrontation. These are *issues* and, except for a minority of cases such as the one about commemorative stamps I just mentioned, there is hardly a matter of policy that is free from controversy and prompts no issue today. The list ranges from nuclear energy and environmental deterioration, through genetic engineering, fetal research, and abortion, to preferential treatment in education and business, and the international debt. Indeed, it continues to grow at an increasing pace, prompting greater numbers of ever more heated discussions, some highly confrontational. These discussions constitute issues in the sense of the term I have developed elsewhere: sharp differences of opinions or conflicts of demands, together with what people do to uphold their opinions or satisfy their demands.[10]

As indicated by the examples just mentioned, there is a great variety of issues, which call for very different kinds of treatment. It is outside the scope of this essay to specify these kinds and establish when they should be used to deal with what issues.[11] Instead, this essay is supposed to describe some general conditions within which all moral problems and issues should be addressed. I will next turn to this task.

MORAL PROBLEMS, MORAL ISSUES, MORAL THEORY, MORAL THEORIES

To say that there are three main kinds of moral problems and many varieties of moral issues does not mean they are not connected with one another. They are. Yet the kinds of problems and varieties of issues are different, and it is a task of moral philosophy to investigate what makes them different. It is also a task of this branch of inquiry to examine whether the kinds of considerations relevant for dealing with moral problems of one kind—say, institutional—are the same as those relevant for dealing with moral problems of another kind—say, of conduct. It is also a task of moral philosophy to examine whether, and, if so, to what extent, in what respects, and in what manner, the kinds of considerations relevant for dealing with given moral problems—say, the problem of whether to have an abortion—are relevant for dealing with related moral issues—say, the problem of whether any and, if so, which, abortions should be permitted by law.

These are but three examples of tasks pursued in doing moral philosophy, or, as it is also called, "ethics as a branch of inquiry." The examples indicate how some of these tasks, though connected with concrete matters of practical concern, are not simply about one or another particular moral problem but about *kinds* of moral problems. They are, accordingly, relatively general in nature and belong at a level of moral philosophy associated with the development of moral theories, a pursuit that individuals, busy pursuing the pressing concerns of their everyday lives, must often put aside.

The help of a sound moral theory in dealing with the moral problems often posed by those daily, as well as related theoretical, concerns may be more than welcome. For the problems characteristically turn out to be complex, difficult, and urgent; and moral theories, the products of moral philosophy, are meant to be instruments for dealing with them.

MORAL THEORIES, MORAL PHILOSOPHY, MORAL ECOLOGIES, AND SOCIAL ECOLOGIES

As one would expect in human affairs, there is far more than one moral theory, and sometimes one moral theory enjoins what another moral theory proscribes. The further theoretical question accordingly arises: Which theory is better and why? And this prompts the higher-level theoretical question: Can one objectively evaluate moral theories? How?

I will address these questions in the ensuing discussion. But, in order to address them soundly, one must, first of all, face up to the fact brought home by the preceding discussion: Moral philosophy and the products of this inquiry, moral philosophies—which, as far as they address kinds of moral problems and not merely particular moral problems taken one by one, amount to moral theories—do not exist in a vacuum. This is true in at least two respects. First, moral philosophy and moral philosophies develop in the midst of *moral ecologies*. These include networks of mores, which, in more developed social groups, are interconnected and interact with laws. Also, in most, if not all, groups, these networks of mores are interconnected and interact with the various morals of individuals and involve both interpersonal influences—say, those among friends and coworkers—and political pressures.

These characteristics of moral ecologies, though not necessarily exhaustive, give an idea of the components of moral ecologies. They are all components of *social ecologies* that the people involved perceive as significant to individuals' lives. In essay 7, social ecologies were characterized as special adaptations and networks of interconnections between social groups, their habitats, and other groups, as well as to special adaptations and interconnections between these groups' individual members, their group or other groups, and their habitats. This should help clarify the ecological nature of the components of moral ecologies. And the specification of these latter components, provided by this essay's preceding discussion, should provide a more detailed account of the components and internal articulation of social ecologies.

To say this is not to say that moral ecologies are identical to social ecologies. For the latter may include adaptations that are not perceived by the people affected as crucial for individuals' lives, maybe because they are not aware of the adaptations at all. For example, a society may practice agriculture in three neighboring valleys simply on the basis of ritualistic grounds. Yet, since weather patterns in the area vary for each valley, the practice may constitute an adaptation that allows the society to have enough food even when the harvest is wiped out in one or two valleys. Given its crucial significance for the lives of the society's members, this practice is a crucial component of the society's *social* ecology. However, given the society's members' ignorance of the practice's survival value, it is *not* part of the society's *moral* ecology. For they do not perceive its significance at all. In other words, though embedded in social ecologies, moral ecologies are not identical to them. Moral ecologies are networks of such things as morals, mores, laws, interpersonal influences, political pressures, and other components of social ecologies that the people involved consider crucial to individuals' lives.

Further, moral ecologies are not identical with ideologies for the same reason ideologies—as argued in essay 10—are not identical with ecologies of belief. Moral ecologies, in contrast with ideologies, range over not merely uneven and conflictive, but also even and nonconflictive, distributions of power and resources.

The second, related respect in which moral philosophy and moral philosophies do not exist in a vacuum concerns the fact that they are prompted by problems posed in the midst of moral ecologies. No doubt, as made plain by this book's discussions so far, moral ecologies involve much that is parochial and oppressive, when not bluntly intolerant and cruel. But the moral philosophies and theories developed in their midst are irrelevant unless—however critically—they address the problems that prompt them. And, in order to address them, they cannot simply ignore the moral ecologies in the midst of which they are posed.

Indeed, a crucial component of the success with which moral philosophy and given moral philosophies help deal with given moral problems is the extent to which they are sensitive to the moral ecologies in which the problems are posed. For if the activity of moral philosophy, and the philosophies formulated in carrying out this activity, are not sensitive to but ignore these ecologies, then moral philosophy is bound to be irrelevant to the problems that prompt it, and moral philosophies' injunctions are unlikely to be feasible or effective in addressing these problems, which, after all, are their reason for being.

Disregarding moral ecologies when engaging in moral philosophy and the evaluation of moral philosophies and theories is both politically unwise and morally objectionable. There are at least two reasons for this. First, it relegates social conflicts and tensions, such as those about the arms race and, after the end of the cold war, about regional hegemony and the proliferation of high-tech weapons to the realm of moral externalities. Indeed, nonwar-related conflicts and tensions about such things as abortion, environmental deterioration, and race relations are also relegated to this realm. But this is a mistake because the said conflicts and tensions are at the center of the problems.

Second, as a result of their narrow view of moral problems, accounts that disregard moral and social ecologies in approaching and dealing with the problems, as a rule, tend to make things even less governable and more conflictive than they already are. For they invariably bring things to a head. In other words, such accounts are not only, as previously argued, untrue to the facts of morality. They are also a sure prescription for catastrophe.

MORAL ECOLOGIES AND THE SOCIAL TESTING OF MORAL HYPOTHESES

There is, however, a line of thought that grants the points just made but emphasizes that, as a matter of morality, we should not enslave our moral decisions to mores and laws that, by and large, are highly questionable. An example of this view can be found in the work of J. L. Mackie, who states:

> The prescription "Think of a set of rules and principles the general adoption of which would best promote what you value and see as worthwhile, and then follow them yourself, regardless of what you think others will do" may well be a recipe for disaster.
>
> The prescription "Think of a set of rules, and try to secure their general acceptance" may be impractical. What the individual can do is . . . to put pressure on some fragments of the system, so that they come gradually to be more favorable to what he sees as valuable and worthwhile.[12]

Mackie's account involves a questionable conception of policy and decision problems. In this conception, for an individual to have a problem of conduct is to have a mere problem of expediency in attaining or furthering a certain aim assumed not to be itself at issue. And for a society to have an institutional problem is for it to have a mere problem of expediency concerning what policies will best lead to attaining certain aims assumed not themselves at issue. That is, it is characteristic of mere problems of expediency that the aims are taken for granted and assumed settled. Mackie's account is a perfect example of this expediency conception of policy and decision problems. For in it what we see as worthwhile or valuable is not at issue, and the problem is simply that of how best to promote it. That is, it is a mere problem of expediency, even if the expediency involved is moral expediency. This, however, leaves Mackie's account open to various objections. Let us consider them next.

Politically Inaccurate

Problems of expediency in social life are not the most frequent or significant ones. Indeed, they have no place among problems concerning such things as abortion, world hunger, environmental decay, and the arms race. In these cases, the aims themselves are characteristically at issue and, often, those involved sharply disagree about what the aims should be. No doubt, some individuals or groups involved approach the issues as if facing a mere matter of expediency in promoting their own aims: what they see as worthwhile or valuable. However, this is merely an element of the higher-level problems centrally posed by the fact that there is controversy and even confrontation about the said aims. It is in no way an accurate conception of these higher-level policy and decision problems.

Politically and Morally Questionable

The said conflicts pose the problem of what to do about the conflicts of aims and the sharp disagreements, and even confrontations, that accompany them. This calls for an approach that is not, as a rule, bound to side with one aim or another. But Mackie's position makes no room for such a thing. For, according to it, one can change aims but, at any one time, one must side with one aim, or set of aims, or another. After all, in his account, each and every individual policymaking action must be aimed at promoting what the individual sees as worthwhile or valuable and as being shared enough by others to have a chance of advancement. Hence, this approach fails to address the political fact of the societally disruptive conflicts that policymakers should squarely address. As a consequence, it cannot help address the institutional moral problems partly posed by the conflicts. This inadequacy for dealing with the actual problems of the political world is not simply a matter of irrelevance or inaccuracy. Given the enormous risks many of the problems involve, it is also a politically unwise and morally objectionable position.[13]

By contrast, focusing on the networks of such things as morals, mores, and, in more complex social groups, also laws, all of them often accompanied by interpersonal influences and political pressures, that is, focus on the moral and social ecologies in the midst of which the problems arise, is free from all the said shortcomings. Indeed, it makes it possible, and likely, that the problems will be addressed, and the moral philosophies and theories used to address the problems will be understood, evaluated, and applied in a morally sensitive and effective manner. Hence our fourth thesis: Moral philosophies and theories are best understood, evaluated, and applied by reference to moral ecologies—to networks of morals, mores, laws, moral injunctions, interpersonal influences, political pressures, and other components of social ecologies that the people involved consider crucial to individuals' lives—and are open to critical scrutiny and interactions in the social testing process.

One might, at this point, find the notion of social testing too vague to be of much help. This is a justified concern. Let us address it next.

THE SOCIAL TESTING OF MORAL HYPOTHESES

The preceding discussion points to an approach in which moral hypotheses—from proposed rules to principles and entire theories—can be significantly confirmed or disconfirmed through a social testing process. I will next outline some of the constraints on social testing and the features social testing thereby must have.

Hard to Find Data

First, a crucial fact of social testing is that, at evaluation time, it is often unfeasible to spell out all significant implications of the action, practice, policy, attitude, trait, theory, or any combination of these, being evaluated. Lack of knowledge or descriptive or predictive ability, when not lack of time, prevents it, even if the details happen to be available for inspection or discovery at the outset. Just think of merely predicting the specific implications of a policy or projecting those of a moral rule for assessing policies about the nearly sixty thousand chemicals in the U.S. market (among which about one thousand are new) every year. In effect, only a few can be monitored. And it is extremely difficult to infer which causes what harms or even what statistical risks are associated with the chemicals. This difficulty is compounded when one considers the myriad other matters relevant for concluding what would be the side effects of given policies regulating the chemicals or the implications of moral rules for assessing the policies.

Unsettled Data

The above information is not merely hard to find. It is not always available. This is characteristically the case when new technologies are involved—for example, the automobile at the beginning and personal computers at the end of the twentieth century. For, at the time of their introduction, some relevant considerations—say, public adaptability to the new technologies—are still unsettled. In any such situation, many a significant detail still has to be worked out, quite often in an unpredictable fashion. This can be done only through a pragmatic process that may include bargaining, negotiation, interaction among those involved, and their limited and only partially controlled experience of the technologies or business or political arrangements under discussion.

For example, as I have discussed at length elsewhere, when automobiles were introduced in the United States at the beginning of the twentieth century, nobody could have predicted the carnage they were going to bring upon the U.S. population. For this carnage was a significant result of the interstate highway system, whose construction was motivated by the cold war. And no data were available then to predict the cold war, hence the interstate highway system. Nor were any data available to predict the fact that the U.S. population was going to find the carnage tolerable enough not to outlaw the automobile or significantly restrict its use. This is not just a matter of unpredictability. A process involving various social decision procedures—most significant, critical scrutiny and social interaction with the new technology and the people affected by it—helps work out morally significant details such as these.[14]

No doubt, from the start, the process involves reasons for morally assessing policies and decisions. For example, upon cool and careful reflection, people may prefer social arrangements that do not preempt future corrections. Say, concerning automobile use, those that would permit corrections when the seriousness of smog and highway fatalities become evident.

In addition, the parties involved may want to leave options open, so that these are worked out through negotiation, bargaining, and other social decision procedures that make room for critical scrutiny and social interaction. For example, they may want to engage in dialogue, negotiation, and bargaining concerning arms reductions. This is not just a matter of gathering further evidence so that one can make more accurate predictions or determine utilities more precisely so that a sounder political judgment can be formed. That is, the utilities themselves, and their very relevance—which may be crucial in such things as arms reductions but secondary or irrelevant in other cases, say, in deciding how to educate one's own children—are worked out in the process, because people's attitudes and judgments about the alternatives are formed and transformed with the process. This process of critical scrutiny and social interaction leads to the settlement of reasons for evaluating such things as actions, practices, policies, a variety of attitudes and traits, and aims concerning one's own personal development or that of others. For example, it leads to new attitudes and judgments about automobile use, and these constitute reasons for evaluating our actions or practices concerning automobile use, and other attitudes we may have concerning automobile use policies.

A Pragmatic Process of Critical Interaction

This social testing process helps work out policy-making details in a manner that involves critical scrutiny and interaction. That is, it relies on judgment. It also relies on individual or collective decisions to interact, and the resulting interaction with such things as new technologies and social arrangements and, through these, with people. In doing so, it leads to the settlement of reasons for further testing moral and related practical hypotheses. For if the details worked out through the process indicated above do not fit the general reasons the hypotheses specify, these hypotheses become questionable. None of these, however, are tested, let alone settled, without critical scrutiny. They become settled when they survive the test—often a recurrent test of critical scrutiny by those going through the critical scrutiny and social interaction process.

SOCIAL TESTING AND GOOD SENSE

No doubt, one group of individuals may settle on what another group would not settle at all. This does not simply depend on what particular concerns are at issue. It also depends on how much care, imagination, moral sensitivity, and political savvy the members of each group have. Some may be so prone to confrontation that nothing will ever be settled among their members. These will simply work themselves into societal extinction (perhaps through argumentative exhaustion). But this is no reason to fault the social testing process. Expecting the process to work regardless of what specific people are involved in it is a mistake. Indeed, it is as hopeless as expecting sound criteria for preventing, diagnosing, and treating disease to have their intended effect when used by careless or ignorant doctors on equally careless or ignorant patients. Though guidelines are helpful and, quite frequently, necessary, they are no substitute for a basic modicum of good sense.

Since social testing is to be carried out in actual social life, it is likely to involve people with various degrees of care, imagination, moral sensitivity, political savvy, and tact. Indeed, some are likely to be very careless, unimaginative, morally insensitive, or lacking in political wisdom or sensitivity to others in both thought and conduct. But this is no obstacle to carrying out the test nor reason to question its validity. So long as, on the aggregate, a basic modicum of good sense is predominant, or shared by few but influential individuals in the group testing a hypothesis, the test should yield reliable results. The further critical scrutiny and trials a hypothesis survives under such circumstances, the more confirmed it is; the less it survives them, the less confirmed it is. If a hypothesis fails more than it survives this testing, then there is reason to seek modifying, supplementing, or substituting it. At no point during or after the testing, however, should one expect total convergence of opinion or coincidence of concerns. Nirvana and perpetual peace are not part of the real world; but reason and basic good sense can find room in this world and through its conflicts.

NOTES

1. For a discussion of this and related points, see Marcus G. Singer, "The Ideal of a Rational Morality," presidential address delivered before the Eighty-fourth Annual Meeting of the Central Division of the American Philosophical Association, St. Louis, Missouri, May 2, 1986, *American Philosophical Association's Proceedings and Addresses* 60, no. 1 (September 1986): 15–38,

2. A brief—and perhaps useful—discussion of acting out of character and related matters can be found in my "Character Traits," Ph.D. thesis, University of Wisconsin-Madison, pp. 17–20, 55–57.

3. *Roe* v. *Wade*, 410 US 113 (1973).

4. For a very useful discussion of custom, law, and morality, and the role of the example just mentioned in the development of a moral order, see Kurt Baier, *The Rational and the Moral Order* (Chicago and LaSalle, Ill.: Open Court, 1995), pp. 210–14.

5. H. L. A. Hart, *The Concept of Law* (Oxford: Oxford University Press, 1961), pp. 89–96.

6. The case described is real, but the names have been changed to respect the privacy of the individuals involved. For a more detailed description and discussion of this case, see my *Contemporary Moral Controversies in Business* (New York and Oxford: Oxford University Press, 1989), pp. 20–21, 90–91.

7. Baier, op. cit., pp. 211–12.

8. As I discussed in previous essays and elsewhere, the question can be addressed objectively in that, as stated, reasons can be offered in support of or against critical judgments of laws or moralities. See essay 8 in this book and essay 4 in my *Philosophy as Diplomacy* (Amherst, N.Y.: Humanity Books, 1994), pp. 40–52.

9. A parallel discussion that focuses on the senses of the term *ethics* can be found in my *Contemporary Moral Controversies in Technology* (New York and London: Oxford University Press, 1987), pp. 3–4, and *Contemporary Moral Controversies in Business* (New York and London: Oxford University Press, 1989), pp. 3–4. As mentioned there, a very succinct discussion of the same topic that draws distinctions akin to those here is included in Marcus G. Singer, ed., *Morals and Values* (New York: Scribner's Sons, 1977), p. 11. In contrast to Singer's distinctions, however, the distinctions I am drawing here do not rely on the notion of a code of conduct but, rather, on the idea of a person's or group's beliefs and presuppositions about right and wrong, good and bad, justified and unjustified. This wider perspective is capable of explaining a code of conduct. Singer provides a more developed version of his distinctions in terms of rules, principles, or standards of conduct involved in ideas of right and wrong in his (previously cited) presidential address. In contrast to Singer's later version, the distinctions devel-

oped in this and my other books are not restricted to problems of conduct and ideas of right and wrong. In addition, they leave open the question of whether all ideas of right and wrong involve rules, principles, or standards.

10. For a discussion of issues and some crucial moral problems they pose, see my *Philosophy as Diplomacy*, passim, especially "Issues and Issue-Overload: A Challenge to Moral Philosophy," pp. 1–12.

11. For a taxonomy of issues and criteria for addressing them, see my *Philosophy as Diplomacy*, "A Delicate Balance: Reason, Social Interaction, Disruption, and Scope in Ethics and Policy Making," pp. 87–108, and "Practical Equity: Dealing with the Varieties of Policy and Decision Problems," pp. 112–44.

12. J. L. Mackie, *Inventing Right and Wrong* (New York: Penguin Books, 1977), p. 148.

13. Here I am focusing on Mackie's conception of policy and decision alone. As for his error theory and how it relates to the matters just discussed, see essay 5 in my *Philosophy as Diplomacy*, especially pp. 62–64.

14. For this example, see my *Philosophy as Diplomacy*, pp. 32–33, 161. The discussion in this section is based on the discussion in *Philosophy as Diplomacy*'s essay 5, especially pp. 64–66.

Dialogue

— I must confess, the overall picture you are painting is quite dizzying. Scientific and philosophical inquiry, which partly overlap each other, are carried out in the midst of also partly overlapping intellectual ecologies—scientific and philosophical ecologies—which, besides the beliefs held in the scientific or philosophical community, also include relevant ordinary beliefs. In addition, these intellectual ecologies—as well as other, nonintellectual ecologies of belief—are embedded in social ecologies, which also include moral, aesthetic, and other ecologies. Have I got it right?

— I believe so, though you seem to suggest that philosophical ecologies are ecologies of belief unrelated to practice.

— I thought that was your view.

— It is not. Philosophical ecologies are ecologies of belief that may or may not be related to practice, depending on what the beliefs are. If the philosophical beliefs concern matters of knowledge, they may not be related to practice. If, however, they concern matters of morals, they belong in moral philosophy, hence, are both prompted by and concerned with practice.

— Things are even more dizzying than I had thought.

— As I said before, I consider this a virtue of my account. For, if it were simpler, it would oversimplify reality, hence, be inadequate.

— I know . . . I know. . . . At any rate, assuming I can get a grasp of all the relationships you have described, I am still quite unclear about the very notion of ecology that you are using.

— My examples should have clarified it to some extent.

— They have; but not enough.

— Do you mean more examples are needed?

— Not just that. More examples accompanied by a more general account of your conception of ecology is needed.

— What do you have in mind?

— I have in mind the beginning of a theory of ecology as it applies to the areas of human activity you have discussed: intercultural relations, business, relations between humans and the biological ecologies in which they exist, ecologies of belief, aesthetic ecologies, and moral ecologies.

— What you have in mind is quite a bit. I am glad you only want the *beginning* of such a theory, because that is all I will have to offer.

— That's fine, so long as what you have in mind is not too little of a beginning.

— What would be enough?

— I think you should address at least the following questions: How does the notion of ecology work in the contexts you discussed? How much of human life does it cover? How does your use of this notion relate to past philosophical practice? Are there any precedents? At any rate, what difference does it make to the way in which we do, or should do, philosophy today?

— You drive a hard bargain. I'll try my best.

Cooperation Among Strangers, Structures of Social Life, and Philosophical Ecologies

TWENTIETH-CENTURY CULTURAL FRAGMENTATION, THE EXPATRIATE EXPERIENCE, AND CROSS-CULTURAL ADAPTATIONS AMONG STRANGERS

The African philosopher K. C. Anyanwu, echoing many others, has said of the twentieth-century's cultural fragmentation from which cross-cultural conflicts arise: "The whole world seems to have become united under the metric system as well as the system of 'hook ups' and 'plug ins,' but the spiritual distance between nations seems to have increased enormously."[1]

Among the manifestations of this fragmentation, he mentions the intransigence of nationalism. As for the factors that increase the fears and tensions concomitant with nationalism and the conflicts it creates, he correctly identifies the denial or lack of discussion of issues concerning cultural experience and values, human dignity and integration, and human coexistence and tolerance.

The discussion of the expatriate experience at this book's outset is a start in the needed direction. For it points to salient features of this experience, which can be used to identify the factors at work in particular cases of cultural fragmentation and conflict and to address the fragmentation and conflicts by appeal to the manner in which expatriates adapt to their foreign cultural environments. Yet, various questions arise: Are the kinds of interpretations, interactions, and convention-settling processes employed by individual expatriates to adapt to their foreign environments also operative at the macrolevel, in interactions between different cultural groups? Are these interpretations, interactions, and convention-settling processes also operant in other relations between individuals and social groups? What social structures and models of social and intellectual development are suited for understanding these processes? In this essay, I will address these questions and offer reasons for three theses. First, because the characteristics of the expatriate experience involve cross-cultural interpretations and interactions, which are at the center of multiculturalism, these characteristics can help us to understand and deal with the cultural fragmentation and cross-cultural conflicts involved in the multiculturalism issue. Second, these characteristics can help by indicating the concerns typically involved in such interpretations and interactions—such as the concern with retaining one's sense of personal identity while attaining one's aims in a foreign environment—as well as the convention-settling processes—for example, as previously mentioned, argumentation, negotiation, quiet resistance, and various forms of outright confrontation—which are used in these interpretations. Third, these and other convention-settling processes can be understood through an ecological model involving a multiplicity of feedback loops, dynamic interactions, and persistent standing conditions, such as policy structures and conceptual frameworks and methods of inquiry and critical scrutiny.

In offering reasons for these theses, I will examine the nature and scope of social cooperation, some of the convention-settling processes it involves, the life structures within which these processes develop, and the dynamic structure of the processes themselves. Let us turn to cooperation by focusing on the type of cases discussed at the book's beginning: those involving the expatriate experience and cross-cultural adaptations.

LARGE-SCALE COOPERATION

The manner in which expatriates adapt to foreign environments involves at least minimal, mutual cooperation between them and the individuals with whom they come in contact when they need directions, or become their roommates, coworkers, or neighbors. Such cooperation among strangers in interactions among individuals is not in question among social scientists and significant sectors of the general public. The problem is that cross-cultural fragmentation and conflicts also take place in society at large, and addressing them would seem to require some form of large-scale societal cooperation. This has led many social scientists and others to focus on economics and the law. But, as David W. Brown says in *Cooperation Among Strangers*, there is some urgency in asking whether large-scale cooperation is possible among strangers without market incentives or government coercion, "because of the growing skepticism about market or government remedies alone to answer to our substantial public problems."[2]

Brown correctly points out that, from everyday conventions, we learn at least three things. First, interdependence leads strangers to be interested in coordination. Second, coordination requires a sort of behavior that is a rudimentary form of social organization. Third, such behavior can produce satisfactory outcomes for problems that no one can resolve alone.[3]

As for the possibility of large-scale cooperation among strangers, there is evidence that it can and does occur. For example, because of cooperation among strangers, water consumption did not have the predicted catastrophic consequences in New York City in 1992. Indeed, this is not an outsider's impression. It was the belief of the New York Department of Environmental Conservation, which appealed to New Yorkers and their experiences by taking full-page ads encouraging New Yorkers to turn off the tap when brushing their teeth, to use washers/dishwashers only with full loads, and to limit showers to three minutes. As for legal and market remedies, the metering schemes that could have been introduced through governmental action were bound to be either unfair and lead to building abandonment by landlords or extraordinarily prohibitive and logistically unworkable for the government.[4]

This is not an isolated example. As Brown mentions, organizational changes such as increasing productivity levels, promoting more minority members to high-ranking positions, and consulting across departments before major decisions are made, evidence cooperation among strangers. To be sure, at one time these practices were unheard of or unattainable. For a time there was confusion and resistance. Yet, eventually, enough people cooperated, some old-timers left, and newcomers took the practices to be the "done thing." Now these conventions are described by everybody as "the way we do things around here."[5] This is an example of a convention-settling process.

The same sort of convention-settling process can be found not merely at the local and organizational levels, but at the level of whole societies. Think of the use of *he* to refer to a person. Not long ago, it was common practice. Using *he or she*, or some equivalent, was unheard of and thought unattainable. Yet, after some resistance, the new convention has become quite general and sufficiently settled. No doubt, conflict was involved; but generalized cooperation among strangers has contributed in a crucial way to the change. The same can be said of the decreased prevalence of a variety of ethnic jokes.

Cooperation among strangers and the convention-settling—which, once conventions are settled, become *reinforcing*—processes it involves, are not restricted to linguistic usage. Think of the mail system. Many of us may take it for granted or think that it works because of the penalties applied to those who, entrusted with mail, neglect to deliver it, or open it. But these penalties by themselves are difficult to apply—at least because the regulations involving them are hard to enforce—in faraway places or in urbanized countries where the legal system is not strong. Yet, the mail generally arrives! And it does so significantly because of customary conventions that involve unenforced large-scale cooperation. Such cooperation is a crucial constituent of large-scale adaptations among strangers.

In fact, the same point can be made regarding the very existence of societies. We take their existence for granted, and some of us may think that they exist because their legal systems make them exist. But legal systems are too limited to be effective without extralegal cooperation among members of society who are total strangers to one another. Just think of the infinitely cumbersome and ineffective bureaucracy that would develop if it were necessary to make sure that most members of society obey the law, that most laws are effectively enforced, that most law enforcers are sufficiently supervised to ensure they enforce the law legally and effectively, that most of those supervising these enforcers are not taking bribes to look the other way, that most of those supervising these supervisors are not doing the same thing, and so on and so forth. This simply would not work, and, on its basis, societies would not exist. But they do, and a good hypothesis explaining this fact is that a great deal of unenforced cooperation through following, settling, and reinforcing conventions among total strangers takes place and helps cement societies together.

The idea suggested here is that the kind of convention-settling processes involved in the expatriate experience also have a large-scale application. So, just as individual expatriates make use of these to build a life for themselves in a foreign culture, so, too, large cultural groups go through the same, or significantly analogous, convention-settling processes in working out mutual adaptations with one another and with individuals who are otherwise strangers. Hence, there is good reason to believe that not just interindividual, but large-scale, cross-cultural adaptations—hence, in the sense of ecologies discussed in preceding essays, large-scale, cross-cultural *ecologies*—can be worked out through the cooperation involved in settling new conventions.

This process, which involves a variety of interactions and interpretations, has traditionally taken place in a largely unintended fashion. However, the enormous increase in cross-cultural interactions in the twentieth century may make it necessary that we become more aware of the process and the ways in which it can take place. At any rate, as argued in previous essays, some salient characteristics of the expatriate experience are ambiguity about one's evoked or hoped-for homeland, a fluid transculturational balance, a heightened awareness of local conventions, and an increased reliance on convention-settling processes. The present essay's first two theses make use of these characteristics: First, because the characteristics of the expatriate experience involve cross-cultural interpretations and interactions, which are at the center of multiculturalism, these characteristics can help us to understand and deal with the cultural fragmentation and crosscultural conflicts involved in the multiculturalism issue. Second, these characteristics can help by indicating the concerns typically involved in such interpretations and interactions—such as the concern with retaining one's sense of personal identity while attaining one's aims in a foreign environment—as well as the convention settling processes—for example, as previously mentioned, argumen-

tation, negotiation, quiet resistance, and various forms of outright confrontation—which are used in these interpretations.

These theses have been partly confirmed by cases discussed in previous
chapters and will be further articulated in this essay's remainder. At this point,
however, I will turn to some of the predominant philosophies involved in the
interactions just described, to the reasons why these philosophies can be said to
constitute philosophical ecologies, to the constraints philosophical ecologies
place on the manner in which a variety of issues should be addressed, and on the
role that philosophy as a branch of inquiry can have in dealing with these issues.

SOCIAL LIFE STRUCTURES
AND PHILOSOPHICAL TRADITIONS

Fiske's Four Models

Human beings relate to each other in an enormous variety of ways. Yet, can
this variety be reduced to a few basic kinds of relations? In his 1991 *Structures of Social Life*, Alan Page Fiske argues that it can: "Whenever human
beings relate to each other, they organize their social relations on the basis of
four elementary psychological models."[6] Fiske's names for these basic models
or forms of social relations are communal sharing, authority ranking,
equality matching, and market pricing. He describes them as follows:

> Communal Sharing is a relation of unity, community, undifferentiated col-
> lective identity, and kindness, typically enacted among close kin. Authority
> Ranking is a relationship of asymmetric differences, commonly exhibited in
> a hierarchical ordering of statuses and precedence, often accompanied by the
> exercise of command and complementary displays of deference and respect.
> Equality Matching is a one-to-one correspondence relationship in which
> people are distinct but equal, as manifested in balanced reciprocity (or tit-
> for-tat revenge), equal share distributions or identical contributions, in-kind
> replacement compensation, and turn taking. Market Pricing is based on an
> (intermodel) metric of value by which people compare different commodi-
> ties and calculate exchange and cost/benefit ratios.[7]

Fiske does not consider these models to be mere logical possibilities. Nor
does he intend to support them with impressionistic discussions alone. As he
states in his epilogue to the 1993 paperback edition, his theory is based on
fieldwork, ethnographic comparisons, and social theory, and it receives independent confirmation from individual cognition studies.[8]

Of particular interest to this book's aims is the fact that Fiske's models
are arguably represented in philosophical traditions. According to Fiske, *com-*

munal sharing, at its core, "is a relationship based on duties and sentiments generating kindness and generosity among people conceived to be of the same kind, especially kin. . . . However, unrelated people may also have a Communal Sharing relationship. People in such a relationship are oriented toward the group, in direct or implicit contrast to outsiders beyond its boundaries."[9] One can arguably find a form of communal sharing involving all humans represented in the thought of David Hume. In his *Enquiry Concerning the Principles of Morals*, for example, Hume writes:

> Sympathy, we shall allow, is much fainter than our concern for ourselves, and sympathy with persons remote from us, much fainter than that with persons near and contiguous; but for this very reason, it is necessary for us, in our calm judgments and discourse concerning the characters of men, to neglect all these differences, and render our sentiments more public and social. Besides that we ourselves often change our situation in this particular, we every day meet with persons, who are in a situation different from us, and who could never converse with us, were we to remain constantly in that position and point of view, which is peculiar to ourselves. The intercourse of sentiments, therefore, in society and conversation, makes us form some general unalterable standard, by which we may approve or disapprove of characters and manners. . . .
>
> Thus, . . . the merit, ascribed to the social virtues, appears still uniform, and arises chiefly from that regard, which the natural sentiment of benevolence engages us to pay to the interests of mankind and society.[10]

Fiske's second model of human social relations is *authority ranking*: This is a relationship of inequality whereby people are ordered in a linear hierarchy and rank, based on social importance or status. It is

> associated with the extent or extension of the self: High-ranking people control more people, things, or land than others, and may be thought to possess more knowledge and mastery over events. . . . Initiative often rests with the highest ranking person or people in a social relationship, and authority typically confers certain related prerogatives involving choice and preference. Attention too is asymmetric, with subordinates less salient than authority figures. . . . Characteristically, inferiors are deferential, loyal, and obedient, giving obeisance and paying homage to their betters.[11]

A version of authority ranking can be found in Friedrich Nietzsche's conception of the social conditions conducive to the enhancement of human beings:

> Every enhancement of the type "man" has so far been the work of an aristocratic society—and it will be so again and again—a society that believes in the long ladder of an order of rank and differences in value between man and man, and that needs slavery in some sense or other.[12]

Fiske characterizes his third model of human social relations, equality matching, as "an egalitarian relationship among peers who are distinct but coequal individuals."[13] There are philosophical examples of this model. Indeed, Fiske suggests the principle of compensatory justice in the context of redressing transgressions and the principle of distributive justice in allocating societal benefits, freedoms, and burdens.[14] An influential version can be found in Immanuel Kant's principle of personality: "Act so that you treat humanity, whether in your own person or in that of another, always as an end and never as a means only."[15]

Finally, Fiske's fourth model, *market pricing*, "is a relationship mediated by values determined by a market system. . . . In a Market Pricing relationship people typically value other people's actions, services, and products according to the rates at which they can be exchanged for other commodities."[16]

Fiske suggests that a version of this fourth model can be found in Adam Smith's *An Inquiry into the Nature and Causes of the Wealth of Nations*. This suggestion is in accordance with received interpretations of Adam Smith's work. However, as Patricia Werhane has convincingly argued, such interpretations are too simplistic. To be sure, Adam Smith did argue that the pursuit of one's private interest pursued through the market need not conflict and may contribute to the public good. But he added that this will happen when economic freedom operates under conditions that include significantly generalized prudence, cooperation, and fair competition, circumscribed by well-defined rules of justice. In any case, even if not exemplified in Adam Smith's views, this model is exemplified in those conceptions formulated in economics, social studies, and philosophy, according to which, cost-benefit analysis plays a crucial role in establishing social and individual values.[17]

Given the societally widespread presence of Fiske's four forms of social relations, it is not surprising that there exist influential philosophical versions of them. These involve not just philosophers' views, but also their societal versions and applications, thus constituting philosophical ecologies in the sense explained in previous essays, and to be further explained in this essay's remainder. As already discussed, these *philosophical ecologies* are embedded in social ecologies. In Fiske's terms, they are associated with forms of social life. Let us next examine the variety of these.

Two Additional Models: Minding One's Own Business and Working Things Out

The philosophies associated with Fiske's models are not the only prevalent philosophical models of human social relations, which contributes to raise a question Fiske himself asked: Are the four basic models just mentioned exhaustive? He writes:

When I wrote this book I half expected that readers would bring to my attention other models that operated in many domains of most cultures. No one has yet done so. The lack of counterexamples reinforces my contention that these models are elementary, and may be the only ones that have the defining features set out in Chapter 8.[18]

Now, coming up with counterexamples to Fiske's account is not just a matter of suggesting possible models. One must, in addition, show that they are widely operant in many societies, a task that requires empirical studies unlikely to be available when Fiske made the above remarks. For such studies would take longer than the three years elapsed between 1991, when Fiske published the book, and 1993, when the epilogue to the paperback edition first appeared. In other words, 1993 was too soon to claim confirmation based on the absence of counterexamples.

This is not to say that Fiske's models are not widely shared across many societies. They are. Nor is it to say that Fiske's account is not valuable. On the contrary. My claim is simply that further cross-cultural studies are needed to confirm or disconfirm the models' exhaustiveness.

Here, I can only suggest some models that might warrant investigation. First, a plausible fifth model appears to be the *minding-one's-own-business* model. For it seems irreducible to the others, and, in addition, it is wide spread, not only in modern, complex societies, but also in such simpler social environments as one finds in the Amazon jungle along the borders between Argentina, Brazil, and Paraguay. Indeed, this area has been chosen as a hideaway by many an outlaw—from ordinary criminals to Third Reich escapees—not only because the jungle is a good hiding place, but also because of the prevalence with which people mind their own business in the area.

A philosophical, refined version of this model can be found in Plato's *Republic*:

> But the truth of the matter was, as it seems, that justice is indeed some thing of this kind, yet not in regard to the doing of one's own business externally, but with regard to that which is within and in the true sense concerns one's self, and the things of one's self. It means that a man . . . should dispose well of what is truly his own, and having first attained to self-mastery and beautiful order within himself, . . . he should then and only then turn to practice.[19]

Another—a sixth—model is *working things out*, in which the guiding idea is none of those in the other models but simply to adapt to the circumstances in a merely pragmatic manner. Again, this model not only appears irreducible to the others but is displayed in a variety of cultural environments—especially in complex and rapidly changing societies—and not always as a way of working out differences between the other models. Sometimes, especially among immigrant groups, where traditional assumptions are often abandoned, the working-things-out model appears predominant— not subservient to the other models. Indeed, in such cases, people eventually tend to abandon the other models in favor of working out adaptations to their new environments.

It is probably not an accident that pragmatism, a philosophy that emphasizes the working out of resolutions to social conflicts through a process of give-and-take, was first formulated in the Western world by Charles Sanders Peirce, William James, John Dewey, and others in the United States in the early 1900s, after an enormous influx of European immigrants had been going on for at least three decades.[20]

Fiske might object that the working-things-out model is characteristic of situations in which his four models are, or would be, used discordantly and that his book focuses on "interactions in which participants understand each other, agree about what models to use, and have mutually compatible motives."[21]

This reply, however, betrays a crucial narrowness in Fiske's approach. For societies, especially complex societies, seem to involve numerous—hence significant—interactions in which participants do not understand one another, do not agree about what models to use, and do not have mutually compatible motives. Since, as Fiske claims, his book's approach does not focus on these significant interactions, there may be—and, as indicated in the preceding discussions concerning crosscultural interactions and interactions in complex and rapidly changing societies, there appear to be—other models (such as "minding one's own business" and "working things out") that, because of its narrow focus on compatible motives and mutual understanding and agreement about models, Fiske's approach is incapable of identifying.

In defense of Fiske's approach, one could argue that the dynamic interactions just described are always subservient to Fiske's four models. Along these lines, one could point out, as Harriet Whitehead does and Fiske acknowledges, that "interactants may—knowingly or unknowingly—use different models to generate a given aspect of their interaction, or to judge it." Also, "people may impose a model on others, making them conform to a model which they would prefer not to follow."[22]

However, the lack of agreement on what models to use in such interactions does not always result from using different models. As indicated above, in cross-cultural interactions involving expatriates who have reached the point of abandoning efforts at applying their traditional models, Fiske's four models need not be used. Instead, the expatriates may simply mind their own business, or try to make things work out in some fashion, *regardless of received models*.

The same appears to be the case in highly complex and rapidly changing societies. Think, for example, of how people tend to interact through the

Internet, say, concerning matters of etiquette. After their attempts at relying on communal sharing, authority ranking, equality matching, or market pricing turn out to be inadequate because, though there is plenty of advice, there are no established ways of doing things concerning Internet etiquette, some cease to rely on traditional models. No doubt, some people tenaciously hold on to one traditional model, or to another, or to some combination of them. But others simply approach the situation with a more open mind and highly pragmatic attitude. In this type of case, strictly speaking, there often is *lack of agreement* rather than disagreement about traditional models, because none of them are advanced. Indeed, if any model is used, it is the working-things-out model.

As stated, the preceding discussion in no way detracts from the fact that Fiske's models are quite valuable and widely used. Yet, they do not fit many dynamic social interactions in highly complex societies well. This should come as no surprise, because the models were developed by focusing on rather static—even when cross-cultural—social interactions in simpler and stable societies. Hence, in order to deal with interactions in the rapidly changing and complex social situations of the present book's concern, we will do well to explore other models, in particular the socioecological model discussed in previous essays. But this model characteristically involves a multiplicity of feedback loops widely understood as closed-loop, circular causal structures. Accordingly, I will next examine feedback thought both in social studies and philosophy as it is relevant to our purposes.

FEEDBACK THOUGHT IN SOCIAL STUDIES AND PHILOSOPHY, AND THIS BOOK'S SOCIOECOLOGICAL MODEL

Feedback-loop concepts in twentieth-century social studies and philosophy have forerunners in a variety of areas of inquiry, from engineering and formal logic, through economics and biology, to sociology, psychology, and philosophy. In some cases, feedback loops appear to have been implicit. Jon Elster, for example, has argued that Descartes implicitly used a feedback-loop concept, in his *Traité d l'Homme*, when explaining how a rational soul in the brain can respond to external disturbances.[23] In other cases, a feedback loop widely understood as a closed-loop, circular causal structure appears to have been explicitly used. When discussing the influence of credit on prices, John Stuart Mill, for example, wrote:

> The inclination of the mercantile public to increase their demand for commodities by making use of all or much of their credit as a purchasing power, depends on their expectation of profit. When there is a general impression that the price of some commodity is likely to rise, from an extra

demand, a short crop, obstructions to importation, or any other cause, there is a disposition among dealers to increase their stocks, in order to profit by the expected price. This disposition tends in itself to produce the effect which it looks forward to, a rise of price: and if the rise is considerable and progressive, other speculators are attracted, who, so long as the price has not begun to fall, are willing to believe that it will continue rising. These, by further purchases, produce a further advance: and thus a rise of price for which there were originally some rational grounds, is often heightened by merely speculative purchases, until it greatly exceeds what the original grounds will justify. After a time this begins to be perceived; the price ceases to rise, and the holders, thinking it time to realize their gains, are anxious to sell. Then the price begins to decline: the holders rush into the market to avoid a still greater loss, and, few being willing to buy in a falling market, the price falls more suddenly than it rose.[24]

In this passage, Mill first explains the effects of speculation on rising prices as a closed, positive feedback loop: A perceived tendency for the price to rise causes the price to rise, which causes a more widespread perceived tendency for the price to rise, which causes the price to rise even more. At this point, Mill also suggests a negative loop at work: Eventually, speculators perceive that the price has become unjustifiably high, they begin to sell, and the price falls.

A somewhat less explicit conception of feedback loops can be found in John Dewey's analysis of a child's reaching for a candle and his criticism of reflex arc explanations of this behavior.

Now if this act, the seeing, stimulates another act, the reaching, it is because both of these acts fall within a larger coordination; ... More specifically, the ability of the hand to do its work will depend ... upon its control, as well as its stimulation, by the act of vision. If the sight did not inhibit as, well as excite the reaching, the latter would be purely indeterminate, it would be for anything or nothing, not for the particular object seen. The reaching, in turn, must both stimulate and control the seeing. The eye must be kept upon the candle if the arm is to do its work.[25]

Dewey's explanation of the effects of the child's perceiving the light on the child's hand as a continuous process can be interpreted to involve at least a closed, positive feedback loop: The child's perception of the light causes the hand to move toward the light, which causes a sustained perception of the light, the hand, and their relative positions, which causes the hand to keep on moving toward the light. In this process, Dewey can also be interpreted to suggest that a negative loop may become active to keep the band on course: If and when the hand, however slightly, strays away from the direction leading to the light, a perception of this fact results, and this perception causes the hand to correct its movement so as to stay closer to the path leading to the light. Hence, Dewey's view appears to involve both positive and negative feedback loops as components of what he describes as a continuous *coordination* process.[26]

There is, however, a significant difference between the loops described in Mill's account and those suggested in Dewey's. The latter aim at a particular specified, known target while the loops on Mill's account do not have fixed, preestablished points at which they begin and end. These represent two different approaches to the role of feedback loops, which were explicitly introduced in social studies during the 1940s and 1950s and are still in vogue. One tends to emphasize particular events, communication, and control. It was initially represented in the works of Norbert Wiener, Kurt Lewin, and Karl Deutsch. The other approach tends to emphasize the role of closed loops in dynamic behavior. It was originally represented in the works of Arnold Tustin, A. W. Phillips, and Herbert Simon.[27]

These approaches were not always clearly distinguished in applying the concept of feedback loop, and, indeed, the concept of feedback loop itself can be, and has been, used in a variety of senses. First, it can be used in the sense in which George P. Richardson uses the term in his *Feedback Thought in Social Science and Systems Theory*: "In the broad sense to stand for all instances of closed-loop, circular causal structure."[28] Second, it can be used in the narrower sense in which Von Bertalaffy used *feedback* to refer only to systems whose behavior aims at preestablished goals—such as thermostats.[29]

Now, even if, following Richardson, one uses the term in its wider sense, it is crucial which approach to *feedback* helps understand a given process. For example, putting aside Von Bertalaffy's semantical concerns, one could still argue, as he did, that "ecological equilibrium is not feedback, as this notion is used with respect to technical devices,"[30] which he understood as driven by a specific, preestablished goal. It is, rather, the nonpreestablished or fixed product of dynamic interactions among a variety of factors.

These are features most characteristic of the socioecological model discussed in previous essays. This model characteristically involves a multiplicity of feedback loops widely understood as closed-loop, circular causal structures. More specific, processes falling under this model are not driven by preestablished particular goals but develop through dynamic interactions among a variety of factors that typically are accompanied by a variety of persistent standing conditions. For example, in *social ecologies*—that is, ecologies encompassing a network of practices, policies, institutions, and ideas predominant in one or more societies—the standing conditions are constituted by practices, institutions, and ideas that are deeply entrenched in the said societies. Promise keeping is one such practice in many, if not all, societies. Art is one such institution. Individual freedom and happiness are often deeply entrenched ideas. And so long as the respective social ecologies persist, changes in them can only happen within the constraints of these standing conditions. By contrast, as discussed in previous essays, changes in these conditions themselves radically alter the social ecologies to the point of doing away with them in favor of either other ecologies or, alternatively, mere social disaggregation.

Let us turn to *intellectual ecologies*—that is, to ecologies that are embedded in social ecologies and encompass a network of concepts, methods, and practices of inquiry, and are predominant among intellectuals and mutually adapted with the predominant ordinary beliefs of one or more societies. Examples of these ecologies are those of science and philosophy. The persistent standing conditions they involve are significantly constituted by concepts, methods, and practices of inquiry that are deeply entrenched among the said intellectuals. Consider twentieth-century science. In it, concepts such as that of efficient cause, methods of empirical investigation such as the method of difference, and the practice and conceptions of freedom of inquiry are deeply entrenched.[31] And if the intellectual ecology of contemporary science is to persist, changes in it can occur only within the constraints of these standing conditions. By contrast, as discussed in previous essays, changes in these conditions themselves radically alter the intellectual ecology of science to the point of doing away with it in favor of either other scientific ecologies or, alternatively, of mere scientific decay.

In addition to scientific ecologies, *philosophical ecologies* constitute another example of intellectual ecologies. This brings us back to this book's beginning and the manner in which philosophy as a branch of inquiry needs to be conceived in order to address problems and issues posed by conflicts between philosophies (or, more precisely, philosophical ecologies) in a sound manner. Also, given the preceding discussion, it raises the question, What, if any, is the role of philosophy as a branch of inquiry concerning philosophical ecologies and, conversely, what, if any, is the role of these ecologies concerning philosophy as a branch of inquiry? I will address this question next.

PHILOSOPHICAL ECOLOGIES AND PHILOSOPHY AS A BRANCH OF INQUIRY

As argued in essay 2, various conceptions of philosophy—in other words, various philosophies about philosophy—can be found competing with one another today. Some philosophers conceive of philosophy as a *research field* left over when the sciences went their independent ways. Others think of philosophy as *the study of large unsolvable problems*.[32] Still others describe philosophy as *an art* "that uses for its medium not stone or paint and canvas or sound, but argument."[33] There are also those who conceive of philosophy as a *dialogue about fundamental problems*. And there are those who think of philosophy as *the task of underlaborers*, with philosophers working with scientists to help them with conceptual difficulties so that they can solve problems in their respective sciences.[34]

However, as I argued in essay 2, from the standpoint of helping us deal with the variety of today's problems and issues—from problems concerning the nature of the universe, through social issues about matters of health care and race relations, to problems regarding the nature of aesthetic value—these conceptions of philosophical practice are too narrow and constrictive. For, first, they do not help us deal with the multiculturalism issue. Moreover, given that many of the additional problems and issues discussed in this book are analogous to the multiculturalism issue in that they involve conceptual, methodological, or customary gaps and conflicts between different groups, the said conceptions of philosophical practice are too narrow and constrictive to help us deal with these problems and issues.

In order to help us deal with the wide range of problems and issues we face, philosophy needs to be conceived in a manner that allows it to engage in and, hence, become sensitive to, the various concerns and ideas actually involved in the said problems and issues. For these same reasons, conceptions of philosophy as the study of large unsolvable problems, as an art, or as a dialogue about fundamental problems are inadequate. They all fall far short of the mark.

The underlaborer conception of philosophy might be thought to escape the criticisms just formulated, because it joins hands with nonphilosophers and, supposedly, helps them address a variety of practical and theoretical problems. But, as following Kai Nielsen, I also argued in essay 2, philosophers' contributions on the underlaborer conception of philosophy—a developed capacity for drawing distinctions, spotting assumptions, digging out unclarities, seeing relationships between propositions, and setting out arguments, perspicuously—do not amount to unique philosophical contributions.

The conception of philosophy I have proposed as suitable for dealing with the enormous range of problems and issues we face today aims at as much comprehensiveness as can be humanly attained or, at least, at integrating the results and activities of other branches of inquiry and human activities in an overall conception of our world. This is not to say that philosophy aims at providing a privileged view of the universe. Nor is it to say that every philosopher should become a generalist instead of focusing on specific areas, such as the philosophy of biology or the philosophy of psychology. It is simply to point out that, as a collective enterprise, philosophy aims at formulating, clarifying, and dealing with problems and issues that concern the nature of reality and mutual relations between beliefs, reasons, values, and norms.

In other words, and more generally, philosophy aims at formulating, clarifying, and dealing with problems that concern the nature and mutual relations of networks of ideas or outlooks that range not merely over the community of philosophers, or even over the wider community of intellectuals, but over society at large. That is, they concern *philosophical ecologies*. This, then, is the critical and creative role of philosophy as a branch of inquiry concerning philosophical ecologies. Let us next turn to the second part of this section's question: What, if any, is the role of philosophical ecologies concerning philosophy as a branch of inquiry?

However disparate, many, if not all, of these ecologies involve deeply entrenched concepts such as those of individual responsibility and the reality-appearance distinction, methods of inquiry such as the reflection on language, and practices such as that of critical scrutiny. These are standing conditions of philosophical ecologies. No doubt, through philosophical inquiry, philosophical ecologies and their associated social ecologies can change and have changed. But, as with other ecologies, these changes can only occur within the constraints of the said standing conditions. In contrast, changes in these conditions themselves—say, ceasing to presuppose a distinction between appearance and reality or ceasing to engage in critical scrutiny of ideas, actions, attitudes, traits, practices, policies, or institutions—radically alter philosophical ecologies to the point of doing away with them and, arguably, with the practice of philosophy as a branch of inquiry itself.

On the conceptions of philosophy as a branch of inquiry and philosophical ecologies previously described, it is not sufficient—indeed it may not always be possible—to integrate the concepts, methods, and practices of different areas of human activity fully. What is crucial—whether in order to attain further integration or to make sound decisions in its absence—is to establish a fruitful *interplay* between different—sometimes sharply divergent—social, scientific, or philosophical ecologies.

This interplay involves, but does not merely—sometimes not even primarily—involve *dialogue*. As with married couples learning to be good partners in a good relationship, or as foreigners learning to live in a foreign culture, it also involves a variety of *interactions*, which help them learn to live well together. This is accomplished, as previously explained, by settling certain conventions through a variety of interactive processes without preestablished particular goals, even when a thorough integration of concepts, methods, or practices is not attained, or attainable. Hence, given this essay's preceding discussion, we have good reason to conclude the essay's last thesis: The convention-settling processes here involved can be well understood through an ecological model that includes a multiplicity of feedback loops, dynamic interactions, and persistent standing conditions, such as policy structures and conceptual frameworks and methods of inquiry and critical scrutiny.

To argue for the changes in philosophical conception and practice I have argued is not to say that basic research in philosophy and the current organization of such research must change. Nor is it to claim that everyone doing philosophy must start doing things differently. Rather, it is to argue that philosophy must undergo some changes—at least some extensions or redirections—as a collective enterprise. Indeed, the needed changes are not just organizational. They are significantly theoretical and include a conceptually integrated formulation, clarification, and evaluation of alternatives for dealing with the variety of problems and issues humans face today. This must be a result of a collective, multioccupational activity—not the privileged subject of any one philosophy researcher, educational practitioner, or policymaker. Nor can it be the privileged subject of any one philosophical school, discipline, profession, or occupation.

The fact that these matters are not the privileged—subject of any such sector provides additional support to the view that the previously indicated interplay is crucial and requires changes in the collective practice of philosophy. Without including all relevant constituencies, the problems are unlikely to be addressed in a manner that is sensitive to all concerns involved.[35]

One might object here that the preceding discussion shows only that, in order not to become hopelessly irrelevant, philosophy and academic disciplines need to join efforts with other disciplines and nonacademic sectors. But the discussion does not show that the public needs such academic efforts. If the problems are real, they will be addressed, by ordinary people anyway. They will do so with or without the help of other-worldly philosophers, aloof academicians, or busy policymakers.

In response to this criticism, one may say that there is good reason to believe that policymakers—however involved in the process—or the general public—however informed it may become about cross-cultural matters—will not have such an abundance of time or even sufficient wisdom to make all other contributions unnecessary. This is where academic—though, of course, not merely academic—efforts such as those I have described can help.

At least because of its long-standing experience with this type of situation, anthropology should play a role in the process. It has a lot to contribute to clarifying the very notion of culture, formulating the distinction between culture and race, understanding such things as kinship systems and ethnocentrism, and helping, through ethnographic studies, to develop a reliable wealth of cases. In this way, it would contribute to free inquiry from ethnocentric, naive, and romantic (however well-meaning) interpretations, such as the notion that Western thought emphasizes reason while non-Western thought is based on tradition. As for history, its role should be central because it has a lot to contribute to placing problems in the context of the times in which they are formulated. And this context is crucial not only for understanding the problems but also for establishing what would count as solutions to them. The philosophical inquiry here envisioned should proceed in this anthropologically and historically informed direction. Such an approach would help us deal more intelligently, and less parochially, with contemporary problems and issues humans face. It also offers us humans hope of finding a way to rise above our fragmented times.[36]

NOTES

1. K. C. Anyanwu, "Cultural Philosophy as a Philosophy of Integration and Tolerance," *International Philosophical Quarterly* 25 (September 1987): 271.

2. David W. Brown, *Cooperation Among Strangers* (Dayton, Ohio: Kettering Foundation, 1992), p. 5. I am very grateful to the late Professor Lewis A. Dexter for bringing this publication to my attention.

3. Ibid., p. 19.

4. Ibid., pp. 22–23.

5. Ibid., p. 24.

6. Alan Page Fiske, preface in *Structures of Social Life* (New York: The Free Press, 1991), p. vii.

7. Ibid.

8. Ibid., p. 410.

9. Ibid., p. 14.

10. David Hume, *Enquiry Concerning the Principles of Morals,* in D. Raphael, *British Moralists 1650–1800, II* (London: Oxford University Press, 1969), p. 74.

11. Fiske, loc. cit.

12. Friedrich Nietzsche, *Beyond Good and Evil* (New York: Vintage Books, 1966), p. 201.

13. Fiske, op. cit., pp. 14–15.

14. Ibid., p. 15.

15. Immanuel Kant, *Foundations of the Metaphysics of Morals*, in Robert Paul Wolff, ed., *Kant: Foundations of the Metaphysics of Morals* (Indianapolis and New York: Bobbs-Merrill, 1969), p. 54.

16. Fiske, loc. cit.

17. Adam Smith, *An Inquiry into the Nature and Causes of the Wealth of Nations* (Oxford: Clarendon Press, 1976), p. 9. See also Patricia Werhane, *Adam Smith and His Legacy for Modern Capitalism* (New York and Oxford: Oxford University Press, 1991), passim. For contemporary examples of the market pricing model in economics, and social studies that overlap with philosophy, see Gary Stanley Becker, *An Economic Analysis of the Family* (Dublin: Economic and Social Research Institute, 1986), *The Economic Approach to Human Behavior* (Chicago: University of Chicago Press, 1976), and *A Treatise on the Family* (Cambridge, Mass.: Harvard University Press, 1991); Gilbert R. Ghez and Gary S. Becker, *The Allocation of Time and Goods over the Life Cycle* (New York: National Bureau of Economic Research, distributed by Columbia University Press, 1975); and A. Freeman, III, et al., *The Economics of Environmental Policy* (New York: John Wiley, 1973), especially p. 23. For a contemporary philosopher's position along these lines, see Tibor R. Machan, *Capitalism and Individualism: Reframing the Argument for a Free Society* (New York: Harvester Wheatsheaf, 1990), and *The Moral Case for a Free Market Economy: A Philosophical Argument* (Lewiston, N.Y.: E. Mellen Press, 1988).

18. Fiske, op. cit., p. 410.

19. Plato, *The Collected Dialogues* (Princeton, N.J.: Princeton University Press, 1973), *Republic* IV, 443 c-e, p. 686.

20. For a discussion of the classical pragmatists and their relations to U.S. history and related U.S. thinkers, see John E. Smith, *The Spirit of American Philosophy* (New York: Oxford University Press, 1966), passim.

21. Fiske, loc. cit.

22. Ibid.

23. Jon Elster, *Leibniz et la formation de l'esprit capitaliste* (Paris: Aubier Montaigne, 1975), p. 57. The passage Elster interprets this way can be found in René Descartes, *Traité de l'Homme*, trans. and commentary by Thomas Steele Hull (Cambridge, Mass.: Harvard University Press, 1972), p. 22. For a discussion of Elster's interpretation, see George P. Richardson, *Feedback Thought in Social Science and Systems Theory* (Philadelphia: University of Pennsylvania Press, 1991), pp. 47–48.

24. John Stuart Mill, *Principles of Political Economy, with Some of Their Applications to Social Philosophy* (New York: D. Appleton, 1883), pp. 45–46. For a discussion of feedback thought in this passage, see Richardson, op. cit., pp. 78–79.

25. John Dewey, "The Reflex Arc Concept in Psychology," *Psychological Review* 3 (1896): 356.

26. For a discussion of Dewey's conception, see Richardson, op. cit., pp. 75–77.

27. See Richardson, op. cit., pp. 93, 341ff. See also Norbert Weiner, *Cybernetics: Or Control and Communication in the Animal and the Machine* (Cambridge, Mass.: MIT Press, 1948), *Extrapolation, Interpolation, and Smoothing of Stationary Time Series* (New York: John Wiley, 1949), and *The Human Use of Human Beings: Cybernetics and Society* (New York: Houghton Mifflin, 1950); Karl W. Deutsch, "Toward a Cybernetic Model of Man and Society, from Some Notes on Research on the Role of Models in the Natural and Social Sciences," *Synthese* 7 (1948): 506–33, *Nationalism and Social Communication* (Cambridge, Mass.: MIT Press, 1966), and *The Nerves of Government* (London: The Free Press of Glencoe and Coltier-Macmillan, 1963); Kurt Lewin, "Feedback Problems of Social Diagnosis and Action, Part II-B of Frontiers in Group Dynamics," *Human Relations* 1 (1947): 147–53, and *Field Theory in Social Science* (New York: Harper and Brothers, 1951); Arnold Tustin, *The Mechanics of Economic Systems* (Cambridge, Mass.: Harvard University Press, 1953); A. W. Phillips, "Mechanical Models in Economic Dynamics," *Econometrica* 17 (1950): 283–305, "Stabilization Policy in a Closed Economy," *Economic Journal* (June 1954): 290–305, and "Stabilization Policy and the Time-Forms of Lagged Responses," *Economic Journal* (June 1957): 265–77; Herbert Simon, *Administrative Behavior: A Study of Decision-Making Processes in Administrative Organization* (New York: Macmillan, 1947), "On the Application of Servomechanism Theory in the Study of Production Control," *Econometrica* 20, no. 2 (1952): 247–68, "Some Strategic Considerations in the Construction of Social Science Models," in P. Lazarsfeld, ed., *Mathematical Thinking in the Social Sciences* (Glencoe, Ill.: Free Press, 1954), *Models of Man* (New York: John Wiley, 1957), "The Architecture of Complexity," *Proceedings of the American Philosophical Society* 106, no. 6 (1962): 467–82, and "Rational Decision Making in Business Organizations," *American Economic Review* 69, no. 4 (1978): 493–513.

28. Richardson, op. cit., p. 123.

29. Von Bertalaffy, "Toward a Physical Theory of Organic Teleology, Feedback, and Dynamics," *Human Biology* 23, no. 6 (1951): 353.

30. Ibid., p. 358.

31. Marjorie Grene, "Puzzled Notes on a Puzzling Profession," *Proceedings and Addresses of the American Philosophical Association*, vol. 61, supp. (Newark, Del.: APA, 1987), p. 75.

32. Ibid., p. 76.

33. Ibid., pp. 76–77.

34. Ibid., pp. 77–78. See also Kai Nielsen, "Philosophy as Critical Theory," *Proceedings*

and Addresses of the American Philosophical Association 61, supp. (Newark, Del.: APA, 1987), pp. 95–99.

35. See, for example, Arthur Caplan, "Can Applied Ethics Be Effective in Health Care, and Should It Strive to Be?" *Ethics* 93 (1983): 311–12. For whatever it is worth, my modest experience in academia—in philosophy, interdisciplinary studies, and the development of interdisciplinary course curricula and workshops—confirms the existence of the conflicts mentioned in the text. But it has also convinced me of the viability of multioccupational approaches and the need for negotiation and other social decision procedures to play a central role in them. For a more detailed discussion of this matter, see my *Philosophy as Diplomacy*, "Bridging Gaps in Babel: Ethics, Technology, and Policy-Making," pp. 181–93; "Like the Phoenix: Ethics, Policy-Making, and the U.S. Nuclear Energy Controversy," pp. 175–78; "A Delicate Balance: Reason, Social Interaction, Disruption, and Scope in Ethics and Policy-Making," pp. 106–108; and "Practical Equity: Dealing with the Varieties of Policy and Decision Problems," pp. 143–44.

In this regard, the conception of philosophy as diplomacy, which I have characterized and substantially discussed elsewhere, should be of help. In a nutshell, philosophy as diplomacy addresses the policy and decision problems posed by issues "in ways that are feasible, effective, and crucially sensitive both to the often unsettled and conflictive nature of the concerns that contribute to pose the problems and to the variety of open-ended social decision procedures that may help settle these concerns and deal with the problems through policies and decisions, and on the basis of reasons worked out in the process." For a more detailed discussion of this conception, see my "Philosophy as Diplomacy," ibid., pp. 70–84. The quoted passage appears on pp. 75–76. Applications of this conception and discussions of how it contrasts with other philosophical approaches can be found throughout the book.

36. This concern has been expressed by anthropologists. See, for example, Richard J. Perry, "Why Do Multiculturalists Ignore Anthropologists?" *The Chronicle of Higher Education*, March 4, 1992, p. A52. For interdisciplinary discussions of culture, see Richard A. Shweder and Robert A. LeVine, *Culture Theory: Essays on Mind, Self, and Emotion* (Cambridge and New York: Cambridge University Press, 1994).

Dialogue

— I know that the approach you suggest aims at being realistic by indicating that philosophy should be comprehensive so far as it is humanly possible. I'm afraid, however, that this will backfire.

— How so?

— In two ways. First, it may lead to more, not less, fragmentation among the various areas of human activity.

— How could it be that bringing people together would lead them to develop less familiarity with and understanding of one another?

— Because, though people would be brought together in cross-cultural, business-related, science- and society-related, and many other areas of dialogue and interaction, these would not be interconnected with one another.

— As a result, you might get more familiarity and understanding among those involved in one or another area, but not among people in one area and people in another area of human activity.

— However, if the people involved in dialogue and interaction concerning one area of human activity—say, business—are from a wide variety of relevant business fields, as well as labor, consumer, and government constituencies, then they are likely to receive roughly the same overall input. Besides, I never said that people involved in dialogue and interaction concerning one area of human activity should not—or do not—try to engage in dialogue and interaction with people involved in other areas of human activity.

— Yes. But would they be more likely to succeed in overcoming frag mentation than under the current circumstances?

— We should not underestimate the fact that the people we are talking about, given the wide variety of walks of life represented in each group engaging in dialogue and interaction, would not only share roughly the same overall input as people in other dialogue and interaction groups, but also engage in critical scrutiny of this input. As a result, their ability to communicate between members of different dialogue and interaction groups would increase. Through this process, fragmentation would be significantly replaced by a common ground thus found or, to some extent, created.

— We have already discussed the nature and scope of the common ground you have in mind. Even if such a common ground is attainable, however, and this is the second way in which I believe your approach will backfire, how will it take the future into account?

— It will to the extent that we care about it and act as trustees of future generations. As I have argued in various essays throughout the book, there is a modicum of common moral sense in human societies, and this secures that enough of us do and will care.

— But do we care about roughly the same things, or is it that some care about the future of society while others care about the future of the planet?

— I have already told you that this way of thinking poses a false dichotomy. In most, if not all, plausible cases, caring about society's future implies caring about the planet's future. Besides, through the dialogue and interaction approach I have outlined, any current differences are likely to be worked out into the common ground we discussed.

— Suppose they do so. In any case, how reliable are our predictions when we are dealing with future generations?

— They become less reliable the further into the future those generations are supposed to exist. However, we can still make some reasonable generalizations about them. They will need at least such things as clean, uncontaminated land, water, and air; fertile soils; predictable weather patterns (hence, a preserved rain forest to regulate them); viable societies; a certain freedom of choice concerning their own lives and future; and a modicum of human companionship.

— These are not minor things. But assuming we can make reasonable generalizations regarding them and future generations, why would they, together with the criteria you formulated, lead to sound practical injunctions and help deal with current fragmentation in human life?

— Throughout the book, I have been giving some reasons to believe that this is the case. However, besides being significantly ecological in the sense I have described and exemplified in the book, my philosophy as diplomacy approach is quite open-ended. In particular, the criteria I formulated should go through a process of social testing, that is, of critical scrutiny and limited interactions with the criteria's implications in human life.

— A question still lingering in my mind is, How is your approach significantly different from what John Dewey advocated as the pragmatic approach to philosophy?

— John Dewey's work has deeply influenced both my *Philosophy as Diplomacy* and the present book. In both works, I have indicated in what regards my approach, though historically related to that of Dewey's, is different from and, I would hope, an improvement on it. Examples of what I consider improvements are the taxonomy of policy and decision problems I used in *Philosophy as Diplomacy*, the hypotheses developed in that book along the lines of the said taxonomy, and the type of feedback processes used in the present book to characterize social change. These and other features are improvements on past pragmatic approaches because they help my approach attain a degree of articulation and specificity that more traditional pragmatic approaches, such as that of Dewey's, lacked. In particular, they help specify the scope and nature of the convention-settling processes that are so significant for cross-cultural and other interactions between individuals and groups with significantly different backgrounds. These, as you know, constitute a salient focus of the present book and have become of crucial planetary concern during the later part of the twentieth century.

— Yet, why would your approach's greater specification and articulation be of more help than traditional pragmatism or, for that matter, than any other approaches?

— This is not a question that can be settled in this book. I have given some reasons for thinking my approach would help. But the decisive test of the approach is significantly a matter of trial by fire. I already made this plain at the end of my *Philosophy as Diplomacy*: As the saying goes, the proof of the pudding is in the eating.

Selected Bibliography

Abbott, A. "Professional Ethics." *American Journal of Sociology* 88 (1983): 855–85.

Ackerman, R. W. *The Social Challenge to Business.* Cambridge, Mass.: Harvard University Press, 1975.

Acton, Jan Paul, and Lloyd S. Dixon. *Superfund and Transaction Costs: The Experience of Insurers and Very Large Industrial Firms.* Santa Monica, Calif.: Rand Corporation, Institute for Civil Justice, 1992.

Addis, Laird. "The Individual and the Marxist Philosophy of History." *Philosophy of Science* 33 (1966): 101–17.

Aftergood, Steven. "The Perils of Government Secrecy." *Issues in Science and Technology* 8, no. 4 (summer 1992): 81–88.

Aiken, William, and Hugh La Follette, eds. *World Hunger and Moral Obligation.* Englewood Cliffs, N.J.: Prentice-Hall, 1977.

Alegría, Fernando, and Jorge Ruffinelli, eds. *Paradise Lost or Gained? The Literature of Hispanic Exile.* Houston: Arte Público Press, 1990.

Al-Issa, Ishan, and Wayne Dennis, eds. *Cross-Cultural Studies of Behavior.* New York: Holt, Rinehart, and Winston, 1970.

Anderson, James E. *Public Policy-Making.* New York: Holt, Rinehart and Winston, 1979.

Anderson, James E., et al. *Public Policy and Politics in America.* Belmont, Calif.: Duxbury/Wadsworth, 1978.

Andrews, Kenneth R. *The Concept of Corporate Strategy.* Homewood, Ill.: Dow Jones-Irwin, 1980.

Andrews, Robert, ed. *The Concise Columbia Dictionary of Quotations.* New York: Columbia University Press, 1989.

Anyanwu, K. C. "Cultural Philosophy as a Philosophy of Integration and Tolerance." *International Philosophical Quarterly* 25 (September 1987): 271–87.

Aram, John D., ed. *Managing Business and Public Policy.* White Plains, N.Y.: Pitman, 1986.

Arrow, Kenneth Joseph. *Social Choice and Individual Values.* 2d, ed. New York: John Wiley and Sons, 1963.

Association of American Publishers, Inc. *Soviet Writers and Journalists in Labor Camp or Internal Exile.* New York: The Association, 1983.

Astrom, Karl J. *Adaptive Control.* Reading, Mass.: Addison-Wesley, 1989.

Atkins, Gary. "In Search of New Objectivity." In Leonard L. Sellers and William L. Rivers, eds. *Mass Media Issues.* Englewood Cliffs, N.J.: Prentice Hall, 1977, pp. 26–34.

Auxter, Thomas. "Toward Multicultural Philosophy." *Teaching Philosophy* 14, no. 2 (June 1991): 187–97.

225

Bachelard, Gaston. *Essai sur la connaissance approchée*. Paris: Vrin, 1969.

Baier, Annette C. *Postures of the Mind*. Minneapolis: University of Minnesota Press, 1985.

―――. *A Progress of Sentiments*. Cambridge, Mass., and London: Harvard University Press, 1991.

Baier, Kurt. *The Moral Point of View*. New York: Random House, 1965.

―――. *The Rational and the Moral Order*. Chicago and LaSalle, Ill.: Open Court, 1995.

Banfield, Edward. *Political Influence*. New York: Free Press, 1961.

Baram, Michael S. "Technology Assessment and Social Control." *Science* 180 (1973): 465–73.

Barber, Benjamin. *The Conquest of Politics*. Princeton, N.J.: Princeton University Press, 1988.

Barbour, Ian G. *Technology, Environment, and Human Values*. New York: Praeger, 1980.

Barry, Brian. *Political Argument*. London: Routledge & Kegan Paul, 1965.

Barry, Brian, and Russell Hardin. *Rational Man and Irrational Society? An Introduction and Sourcebook*. Beverly Hills, Calif.: Sage, 1982.

Barth, H. *Truth and Ideology*. Los Angeles: University of California Press, 1977.

Baum, Robert J., and Albert Flores, eds. *Ethical Problems in Engineering*. New York: Center for the Study of Human Dimensions of Science and Technology, 1978.

Bazelon, David L. "Governing Technology: Values, Choices, and Scientific Progress." *Technology and Society* 5 (1983): 17–18.

Beardsley, Monroe C. "In Defense of Aesthetic Value." In *Proceedings and Addresses of the American Philosophical Association*. Newark, Del.: APA, 1979, pp. 723–49.

Beauchamp, Tom L., and Norman Bowie, eds. *Ethical Theory and Business*. 3d ed. Englewood Cliffs, N.J.: Prentice-Hall, 1988.

Becker, Gary Stanley. *An Economic Analysis of the Family*. Dublin: Economic and Social Research Institute, 1986.

―――. *A Treatise on the Family*. Cambridge, Mass.: Harvard University Press, 1991.

Becker, Gary S., and Gilbert R. Ghez. *The Allocation of Time and Goods over the Life Cycle*. New York: National Bureau of Economic Research, distributed by Columbia University Press, 1975.

Becker, Lawrence C. "The Finality of Moral Judgments." *Philosophical Review* 82 (1973): 364–70.

Beiner, Ronald. *Political Judgment*. Chicago: University of Chicago Press, 1983.

Bentham, Jeremy, "The Principles of Morals and Legislation." In John Stuart Mill, *Utilitarianism, On Liberty, Essay on Bentham, Together with Selected Writings of Jeremy Bentham and John Austin*, ed. Mary Warnock. Cleveland and New York: Meridian, 1968.

Berlin, Sir Isaiah. "On the Pursuit of the Ideal." Turin, Italy: Fondazione Giovanni Agnelli, 1988, pp. 1–16.

Bernard, Thomas J. *The Consensus-Conflict Debate*. New York: Columbia University Press, 1983.

Berry, Wendell. *The Unsettling of America: Culture and Agriculture*. New York: Avon, 1977.

Besnard, P. *L'anomie*. Paris: Presses Universitaires de France, 1987.

Bhushan, Nalini. "The Real Challenge of Cultural Diversity: Clarifying the Boundaries of Legitimate Philosophical Practice." *Teaching Philosophy* 14 (June 1991): 165–78.

Bierbaum, Rosina, and Robert M. Friedman. "The Road to Reduced Carbon Emissions." *Issues in Science and Technology* (winter 1991–92): 58–65.

Bluestone, Barry, and Bennet Harrison. *The Deindustrialization of America*. New York: Basic Books, 1982.

Boas, Marie. *The Scientific Renaissance: 1450–1630*. New York: Harper & Row, 1962.

Bookchin, Murray. *Toward an Ecological Society*. Montreal and New York: Black Rose Books, 1980.

————. *The Philosophy of Social Ecology: Essays in Dialectical Naturalism.* Montreal and New York: Black Rose Books, 1990.

————. *The Ecology of Freedom: The Emergence and Dissolution of Hierarchy.* Montreal and New York: Black Rose Books, 1991.

————. *Urbanization Without Cities: The Rise and Decline of Citizenship.* Montreal and New York: Black Rose Books, 1992.

————. *From Urbanization to Cities: Toward a New Politics of Citizenship.* London: Cassell, 1995.

Borges, Jorge Luis. *Labyrinths.* New York: New Directions, 1962.

————. *Seven Nights.* New York: New Directions, 1984.

Bosselman, F., et al. *The Taking Issue.* Washington, D.C.: U.S. Government Printing Office, 1971.

Bottommore, Tom. *Crisis and Contention in Sociology.* Beverly Hills and London: Sage, 1975.

Bouwsma, William J. "Politics in the Age of the Renaissance." In Joseph Rotschild, David Sidorsky, and Bernard Wishy, eds. *Chapters in Western Civilization.* New York: Columbia University Press, 1961, pp. 199–244.

Bower, Bruce. "Spanish Survives Bilingual Challenge." *Science News* 146 (September 3, 1994): 148.

————. "Return to the Group." *Science News* 148 (November 18, 1995): 328–30.

————. "Ultrasocial Darwinism." *Science News* 148 (November 25, 1995): 366–67.

————. "Rational Mind Designs: Research into the Ecology of Thought Treads on Contested Terrain." *Science News* 150 (July 13, 1996): 24–25.

Bozeman, Barry. *Public Management and Policy Analysis.* New York: St. Martin's Press, 1979.

Brandt, Willy, et al. *North-South: A Programme for Survival.* Cambridge, Mass.: MIT Press, 1980.

Braybrooke, David. *Studies in Moral Philosophy.* Oxford: Blackwell, 1968.

————. *Three Tests for Democracy: Personal Rights, Human Welfare, Collective Preference.* New York: Random House, 1968.

————. *Traffic Congestion Goes Through the Issue-Machine.* London and Boston: Routledge & Kegan Paul, 1974.

————. "Policy Formation with Issue-Processing and Transformation of Issues." In G. A. Hooker, J. J. Leach, and E. F. McClennen, eds. *Foundations and Applications of Decision Theory.* Dortrecht, Holland: Reidel, 1978, pp. 1–14.

————. "Work: A Cultural Ideal Ever More in Jeopardy." In Peter A. French, Theodore E. Uehling Jr., and Howard K. Wettstein, eds. *Midwest Studies in Philosophy VII.* Minneapolis: University of Minnesota Press, 1982, pp. 321–41.

————. *Meeting Needs.* Princeton, N.J.: Princeton University Press, 1987.

Braybrooke, David, ed. *Ethics in the World of Business.* Totowa, N.J.: Rowman & Allanheld, 1983.

Braybrooke, David, and Charles E. Lindblom. *A Strategy of Decision.* New York: Free Press, 1963.

Brechner, K. C. "An Experimental Analysis of Social Traps." *Journal of Experimental Social Psychology* 13 (1977): 552–64.

Breyer, Stephen. *Breaking the Vicious Circle: Towards Effective Risk Regulation.* Cambridge, Mass., and London: Harvard University Press, 1993.

Britt, Philip. "FHFB Will Aim to Boost Community Lending Activity." *Savings Institutions* (July 1992): 9.

Bromberg, Joan Lisa. *Fusion: Science, Politics, and the Invention of a New Energy Source.* Cambridge, Mass.: MIT, 1982.

Bromley, D. Allan. "The Making of a Greenhouse Policy." *Issues in Science and Technology* (fall 1990): 55–61.

Brown, David W. *Cooperation Among Strangers*. Dayton, Ohio: Kettering Foundation, 1992.

Brucker, Herbert. "What's Wrong with Objectivity." *Saturday Review*, October 11, 1969, pp. 77–79.

Buchanan, Allen. *Ethics, Efficiency, and the Market*. Totowa, N.J.: Rowman & Allanheld, 1985.

Buchholz, Eugene, ed. *Business Environments and Public Policy*. Englewood Cliffs, N.J.: Prentice-Hall, 1986.

Bunge, Mario. *Racionalidad y realismo*. Madrid: Alianza Universidad, 1985.

———. *Seudociencia e ideologia*. Madrid: Alianza Editorial, 1985.

Burke, Edmund. *The Works of Edmund Burke*. Vol. 5. Boston: Little, Brown, 1881.

Burke, Kenneth. *Attitudes Toward History*. Los Altos, Calif.: Hermes, 1959.

Burkhardt, Jakob. *Die Kultur der Renaissance in Italien*. N.p., 1868.

———. *Griechische Kulturgesichte*. Vols. 1–5. N.p., 1898–1902.

Busch, Lawrence, and William B. Lacy. *Science, Agriculture, and the Politics of Research*. Boulder, Colo.: Westview, 1983.

Butler, John C., III. "Superfund Super Costs." In *Rethinking Superfund: It Costs Too Much; It's Unfair; It Must Be Fixed*. Washington, D.C.: National Legal Center for the Public Interest, 1991, pp. 67–77.

Byrnes, C. I., and A. Kurzhanski, eds. *Modelling and Adaptive Control*. Berlin and New York: Springer-Verlag, 1988.

Cantor, Geoffrey. *Michael Faraday: Sandemanian and Scientist: A Study of Science and Religion in the Nineteenth Century*. London: St. Martin's Press, 1991.

Caplan, Arthur L. "Ethical Engineers Need Not Apply: The State of Applied Ethics Today." *Science, Technology, and Human Values* 6, no. 33 (fall 1980): 24–32.

———. "Can Applied Ethics Be Effective in Health Care, and Should It Strive to Be?" *Ethics* 93 (1983): 311–19.

Carisnaes, W. *The Concept of Ideology and Political Science*. Westport, Conn.: Greenwood Press, 1981.

Carr, Albert. "Can an Executive Afford a Conscience?" *Harvard Business Review* (July–August 1970): 58–64.

Castro, Donald S. *The Argentine Tango as Social History*. Lewiston, N.Y.: E. Mellen Press, 1991.

Chalk, Rosemary. "Commentary on the NSA Report." *Science, Technology, and Human Values* 8, no. 1 (winter 1983): 21–24.

Chamberlain, N. W. *The Limits of Corporate Responsibility*. New York: Basic Books, 1973.

Churchill, Robert Paul, ed. *Crossing Cultural Boundaries*. Amherst, N.Y.: Humanity Books, forthcoming.

Churchman, C. West. *The Design of Inquiring Systems*. New York: Basic Books, 1971.

Clifford, W. K. "The Ethics of Belief." In *Lectures and Essays*. London: Macmillan, 1879, pp. 177–211; originally published in *Contemporary Review*, January 1877.

Coburn, Alexander, and James Ridgeway, eds. *Political Ecology*. New York: Times Books, 1979.

Coburn, R. "Technology Assessment, Human Good, and Freedom." In K. E. Goodpaster and K. M. Sayre, eds. *Ethics & Problems of the 21st Century*. Notre Dame, Ind.: Notre Dame University Press, 1979, pp. 106–21.

Cohen, I. Bernard. *The Birth of a New Physics*. Garden City, N.Y.: Anchor Books, 1960.

Community Investment Corporation Demonstration Act of 1992. U.S. *Code*, vol. 22, secs. 5305ff. (1994).

Community Reinvestment Act. U.S. Code, vol. 4, secs. 2901ff. (1988).

Connolly, William E. *The Terms of Political Discourse.* 2d ed. Princeton, N.J.: Princeton University Press, 1983.

Cooke, Robert Allan. "Business Ethics at the Crossroads." *Journal of Business Ethics* 5, no. 3 (June 1986): 259–63.

Cowart, David. *Literary Symbiosis: The Reconfigured Text in Twentieth Century Writing.* Athens: University of Georgia Press, 1993.

Cropsey, Joseph. *Political Philosophy and the Issues of Politics.* Chicago: University of Chicago Press, 1977.

Cross, J. G., and M. J. Guyer. *Social Traps.* Ann Arbor: University of Michigan Press, 1980.

Dahrendorf, Ralf. "Out of Utopia: Toward a Reorientation of Sociological Analysis." *American Journal of Sociology* (September 1958): 115–27.

D'Andrade, Roy G. "Cultural Meaning Systems." In Richard A. Shweder and Robert A. LeVine. *Culture Theory: Essays on Mind, Self, and Emotion.* Cambridge and New York: Cambridge University Press, 1984, pp. 88–119.

Daniels, Norman. *Just Health Care.* Cambridge: Cambridge University Press, 1985.

Da Vinci, Leonardo. *Breviarios.* Bueños Aires: Editorial Schapire, 1952.

Dawkins, Richard. *The Selfish Gene.* New York and London: Oxford University Press, 1976.

De George, Richard T. *Business Ethics.* 2d ed. New York: Macmillan, 1986.

Dennis, Wayne, ed. *Readings in the History of Psychology.* New York: Appleton-Century-Crofts, 1948.

De Tracy, Antoine Destutt. *Elements d'Ideologie.* Discussed in David McLellan, *Ideology.* Minneapolis: University of Minnesota Press, 1986.

Der Derian, James. *On Diplomacy.* Oxford: Basil Blackwell, 1987.

Descartes, René. *Traité de l'Homme.* Trans. and commentary by Thomas Steele Hull. Cambridge, Mass.: Harvard University Press, 1972.

Deutsch, Karl W. "Toward a Cybernetic Model of Man and Society, from Some Notes on Research on the Role of Models in the Natural and Social Sciences." *Synthese* 7 (1948): 506–33.

———. *Nationalism and Social Communication.* Cambridge, Mass.: MIT Press, 1966.

———. *The Nerves of Government.* London: The Free Press of Glencoe and Collier-Macmillan, 1963.

Dewey, John. "The Reflex Arc Concept in Psychology." *Psychological Review* 3 (1896): 357–70. Reprinted in Dennis Wayne, *Readings in the History of Psychology.* New York: Appleton-Century-Crofts, 1948, pp. 355–65.

———. *Reconstruction in Philosophy.* Boston: Beacon Press, 1957.

———. *The Quest for Certainty.* New York: Putnam, 1929; Capricorn, 1960.

Dewey, John, and James H. Tufts. *Ethics.* New York: Henry Holt, 1913.

Dexter, Lewis Anthony. "Intra-Agency Politics: Conflict and Contravention in Administrative Entities." *Journal of Theoretical Politics* 2 (1990): 151–72.

Dexter, Lewis Anthony, and David Manning. *White People, Society, and Mass Communications.* New York: Free Press, 1964.

Dolbeare, Kenneth. *Public Policy Evaluation.* Beverly Hills and London: Sage, 1975.

Donaldson, Thomas, and Patricia H. Werhane, eds. *Ethical Issues in Business.* Englewood Cliffs, N.J.: Prentice-Hall, 1983.

Dorf, R. C. *Technology and Society.* San Francisco: Boyd and Fraser, 1974.

Dornbusch, Rudiger. *Open Economy Macro-Economics.* New York: Basic Books, 1980.

Dorner, Peter, and Mahmoud A. El-Shafie. *Resources and Development: Natural Resource Policies and Economic Development in an Interdependent World.* Madison: University of Wisconsin Press, 1980.

Douglas, A. E. *Symbiotic Interactions*. Oxford and New York: Oxford University Press, 1994.

Douglas, Mary, and Aaron Wildavsky. *Risk and Culture*. Berkeley, Los Angeles, and London: University of California Press, 1982.

Drucker, P. F. *Management: Tasks, Responsibilities, Practices*. New York: Harper & Row, 1973.

Dryzek, John S. *Discursive Democracy: Politics, Policy, and Political Science*. New York and Cambridge: Cambridge University Press, 1990.

Dunn, John. *Rethinking Modern Political Theory*. London: Cambridge University Press, 1985.

Dunn, William N. *Public Policy Analysis*. Englewood Cliffs, N.J.: Prentice Hall, 1981.

Dworkin, Ronald. *Taking Rights Seriously*. Cambridge, Mass.: Harvard University Press, 1977.

———. *A Matter of Principle*. Cambridge, Mass.: Harvard University Press, 1985.

Dye, Thomas R. *Understanding Public Policy*. Englewood Cliffs, N.J.: Prentice-Hall, 1972.

Dye, T. R., and V. Gray. *The Determinants of Public Policy*. Lexington, Mass.: Lexington Books, 1980.

Edgerton, Gary R. *Film and the Arts in Symbiosis: A Resource Guide*. Westport, Conn.: Greenwood Press, 1988.

Ellul, Jacques. *The Technological Society*. New York: Random House, 1964.

Elster, Jon. *Leibniz et la formation de l'esprit capitaliste*. Paris: Aubier Montaigne, 1975.

———. *Explaining Technical Change*. Cambridge: Cambridge University Press, 1983.

———. *The Cement of Society*. Cambridge and New York: Cambridge University Press, 1989.

Elster, Jon, ed. *Rational Choice*. New York: New York University Press, 1986.

Elster, Jon, and Aanund Hylland, eds. *Foundations of Social Choice Theory*. Cambridge: Cambridge University Press, 1986.

Enthoven, Alain C., and K. Wayne Smith. *How Much Is Enough?* New York: Harper & Row, 1971.

Equal Credit Opportunity Act of 1975. *U.S. Code*, vol. 6, secs. 1691ff. (1994).

Etzioni, Amitai. *The Moral Dimension: Toward a New Economics*. New York: Free Press, 1988.

Fain, Haskell. *Normative Politics and the Community of Nations*. Philadelphia: Temple University Press, 1987.

Faraday, Michael. "Historical Sketch of Electromagnetism." *Annals of Philosophy* 2 (1821): 195–200, 274–90; vol. 3 (September 1821–February 1822): 107–21.

———. "On Some New Electro-Magnetical Motions, and on the Theory of Magnetism." *Quarterly Journal of Science* 12 (October 1821): 74–96.

———. *Diary*. Edited by Thomas Martin. London: Bell, 1932–1936.

———. *Experimental Researches in Electricity*. New York: Dover, 1965.

———. *The Correspondence of Michael Faraday*. Edited by Frank A. J. L. James. London: Institution of Electrical Engineers, 1991.

———. *Experimental Researches in Chemistry and Physics*. London and New York: Taylor & Francis, 1991.

Faris, Ralph M. *Crisis and Consciousness*. Amsterdam: B. R. Grijbner, 1977.

Federal Financial Institutions Examination Council. *A Citizen's Guide to CRA*. Washington, D.C.: Federal Financial Institutions Examination Council, 1985.

Federal Reserve Board. "Order Denying Acquisition of a Bank." February 15, 1989. Reprinted in Paul H. Schieber and Dennis Replansky. Appendix 2 in *The Lender's Guide to Consumer Compliance and Anti-Discrimination Laws*. Chicago: Probus, 1991, pp. 211–27.

Federal Reserve System, Department of the Treasury, Office of the Comptroller of the Currency, Federal Home Loan Bank Board, and Federal Deposit Insurance Corporation. "Statement of the Federal Financial Supervisory Agencies Regarding the Community Reinvestment Act." *Federal Register* 54, no. 64 (April 5, 1989): 13742–46.

Feuer, L. *Ideology and Ideologists*. Oxford: Blackwell, 1975.

Financial Institutions Reform, Recovery and Enforcement Act of 1989. U.S. Code. Secs. 1790ff.

Fink, Conrad C. *Media Ethics.* New York: McGraw-Hill, 1988.

Fishbein, Allen J. "Banks Giving Credit with a Conscience." *Business and Society Review* (winter 1989): 33–37.

Fisher, Frank. *Politics, Values, and Public Policy: The Problem of Methodology.* Boulder, Colo.: Westview Press, 1980.

Fishkin, James S. *Tyranny and Legitimacy: A Critique of Political Theories.* Baltimore and London: Johns Hopkins University Press, 1979.

Fiske, Alan Page. *Structures of Social Life.* New York: Free Press, 1991.

Foley, Gerald, and Charlotte Nassim. *The Energy Question.* New York: Penguin Books, 1976.

Foreman, John C., and J. McDuffie Brunson. "CRA: Bankers on the DefensiveBankers Focus on Documentation of Community Lending Efforts in Response to New CRA Rules." *Bank Management* (January 1991): 34–37, 40.

Fox, Richard G. *Recapturing Anthropology: Working in the Present.* Santa Fe, N. Mex.: School of American Research, 1991.

Freedman, J. O. *Crisis and Legitimacy.* Cambridge: Cambridge University Press, 1978.

Freeman, A., R. Haveman, and A. Kneese. *The Economics of Environmental Policy.* Baltimore: Resources for the Future and Johns Hopkins University Press, 1973.

French, Peter A., Theodore E. Uehling Jr., and Howard K. Wettstein, eds. *Midwest Studies in Philosophy.* Minneapolis: University of Minnesota Press, 1982.

Friedrich, Carl J. *Authority.* Cambridge, Mass.: Harvard University Press, 1958.

———. *Community.* New York: Liberal Arts Press, 1959.

———. *Liberty.* New York: Atherton Press, 1962.

———. *The Public Interest.* New York: Atherton Press, 1962.

———. *Justice.* New York: Atherton Press, 1963.

———. *Rational Decision.* New York: Atherton Press, 1964.

Fries, Sylvia Doughty. "Expertise Against Politics: Technology as Ideology on Capitol Hill, 1966–1972." *Science, Technology, and Human Values* 8, no. 2 (spring 1983): 6–15.

Frohock, Fred M. *Public Policy: Scope and Logic.* Englewood Cliffs, N.J.: Prentice Hall, 1979.

Fullinwider, Robert. "Multicultural Education." *Report from the Institute for Philosophy and Public Policy* 11, no. 3 (summer 1991): 12–14.

Gardels, Nathan. "Two Concepts of Nationalism: An Interview with Isaiah Berlin." *The New York Review of Books* 38, no. 19 (November 21, 1991): 19–23.

Gassler, Robert Scott. *The Economics of Nonprofit Enterprise: A Study in Applied Economic Theory.* Lanham, Md.: University Press of America, 1986.

Geertz, Clifford. *Ideology and Discontent.* New York: D. Apter, 1964.

———. *The Interpretation of Cultures: Selected Essays.* New York: Basic Books, 1973.

———. *Local Knowledge: Further Essays in Interpretive Anthropology.* New York: Basic Books, 1983.

———. *After the Fact: Two Countries, Four Decades, One Anthropologist.* Cambridge, Mass.: Harvard University Press, 1995.

Gelman, Jorge. "Nuevas Imágenes de un Mundo Rural: La campaña rioplatense antes de 1810." *Ciencia Hoy* 1 (December 1989/January 1990): 57–61.

Gillespie, Norman C. "The Business of Ethics." *University of Michigan Business Review* 26, no. 6 (November 1975): 1–4.

Glick Schiller, Nina, Linda Basch, and Cristina Szanton. *Towards a Transnational Perspective on Migration: Race, Class, Ethnicity, and Nationalism Reconsidered.* New York: New York Academy of Sciences, 1992.

Gobello, José. *Diccionario Lunfardo.* Bueños Aires: A. Peña Lillo, 1975.

Goffman, Erving. *The Presentation of Self in Everyday Life*. Garden City, N.Y.: Doubleday, 1959.

Goodin, Robert E. *Manipulatory Politics*. New Haven, Conn., and London: Yale University Press, 1980.

———. "No Moral Nukes." *Ethics* 90 (April 1980): 417–49.

Goodpaster, K. E., and K. M. Sayre, eds. *Ethics & Problems of the 21st Century*. Notre Dame, Ind.: Notre Dame University Press, 1979.

Goodpaster, Kenneth E., and John B. Matthews Jr. "Can a Corporation Have a Conscience?" *Harvard Business Review* (January–February 1982): 132–41.

Gregg, Phillip M. *Problems of Theory in Policy Analysis*. Lexington, Mass.: Lexington Books, 1976.

Grene, Marjorie. "Puzzled Notes on a Puzzling Profession." In *Proceedings and Addresses of the American Philosophical Association*. Vol. 61. supp. Newark, Del.: APA, 1987, pp. 75–80.

Grubel, Herbert C. *International Economics*. Homewood, Ill.: Irwin, 1981.

Grumm, John G., and Stephen L. Wasby, eds. *The Analysis of Policy Impact*. Lexington, Mass.: Lexington Books, 1981.

Guha, Ramachandra, ed. *Social Ecology*. Delhi and New York: Oxford University Press, 1994.

Habermas, Jurgen. *Knowledge and Human Interests*. Boston: Beacon Press, 1968.

———. *Theory and Practice*. Boston: Beacon Press, 1973.

———. *Legitimation Crisis*. Boston: Beacon Press, 1975.

———. "Morality and the Law." In Sterling McMurrin, ed., *The Tanner Lectures on Human Values, VIII*. Salt Lake City, Utah, and Cambridge: University of Utah Press and Cambridge University Press, 1988.

Hahn, Frank, and Martin Hollis, eds. *Philosophy and Economic Theory*. Oxford and New York: Oxford University Press, 1979.

Halperin Donghi, Tulio. "Argentina's Unmastered Past." *Latin American Research Review* 23, no. 2 (1988): 3–24.

Hamerton, Phillip Gilbert. *The Intellectual Life*. Boston: Roberts Brothers, 1875.

Hamlett, Patrick W. "A Typology of Decisionmaking in the U.S. Congress." *Technology and Human Values* 8, no. 2 (spring 1983): 33–40.

———. "Science, Technology, and Liberal Democratic Theory." *Technology in Society* 6 (1984): 249–62.

Hampshire, Stuart. *Morality and Conflict*. Cambridge, Mass.: Harvard University Press, 1983.

Hardt, J. P., and G. D. Holliday. *U.S.-Soviet Commercial Relations: The Interplay of Economics, Technology Transfer and Diplomacy*. Washington, D.C.: U.S. Government Printing Office, 1973.

Harrington, Joseph, ed. "Credit Policies and Outreach Efforts Get Rigorous CRA Review." *Savings Institutions*, October 1990, pp. 9, 11.

Harris, Eddy L. *Native Stranger: A Black American's Journey into the Heart of Africa*. New York: Simon & Schuster, 1992.

Harris, James. *Against Relativism: A Philosophical Defense of Method*. La Salle, Ill.: Open Court, 1992.

Hart, H. L. A. *The Concept of Law*. Oxford: Oxford University Press, 1961.

Hausman, Daniel M., ed. *The Philosophy of Economics*. Cambridge and New York: Cambridge University Press, 1984.

Hawkes, Jacquetta, Sir Leonard Wooley, et al., eds. *Historia de La Humanidad*. Bueños Aires: Editorial Sudamericana, 1963.

Hay, William Henry. "Under the Blue Dome of the Heavens." Presidential address delivered before the Seventy-third Annual Western Meeting of the American Philosophical Association in Chicago, April 25, 1975. *Proceedings and Addresses of the American Philosophical Association*. Vol. 48. Newark, Del.: APA, 1974–1975, pp. 54–67.

Haynes, Richard, and Ray Lanier, eds. *Agriculture, Change, and Human Values*. Vols. 1 and 2. Gainesville: University of Florida Humanities and Agriculture Program, 1983).

Hayward, Tim. *Ecological Thought: An Introduction*. Cambridge, England: Polity Press; Cambridge, Mass.: Blackwell Publishers, 1995.

Heath, Anthony. *Rational Choice and Social Exchange*. Cambridge: Cambridge University Press, 1976.

Heidegger, Martin. *Basic Writings*. New York: Harper and Row, 1977.

Heilbroner, Robert L. *The Making of Economic Society*. Englewood Cliffs, N.J.: Prentice-Hall, 1962.

———. *Between Capitalism and Socialism*. New York: Vintage, 1970.

Held, Virginia. *The Public Interest and Individual Interest*. New York: Basic Books, 1970.

———. "The Political 'Testing' of Moral Theories." *Midwest Studies in Philosophy* 7 (1982): 343–63.

Held, Virginia, Sidney Morgenbesser, and Thomas Nagel, eds. *Philosophy, Morality, and International Affairs*. New York and London: Oxford University Press, 1974.

Heller, Scott. "Worldwide 'Diaspora' of Peoples Poses New Challenges for Scholars." *The Chronicle of Higher Education*, June 3, 1992: A7–9.

Herber, Lewis. *Our Synthetic Environment*. New York: Knopf, 1962.

Herrick, Charles N. "Science and Climate Policy: A History Lesson." *Issues in Science and Technology* (winter 1991–92): 56–57.

Herskovits, Melville J. *Acculturation*. Gloucester, Mass.: Peter Smith, 1958.

Hobbes, Thomas. *Leviathan*. In D. D. Raphael, ed. *British Moralists 1650–1880*. Oxford: Clarendon Press, 1969.

Hoffman, Ross J. S., and Paul Levack, eds. *Burke's Politics*. New York: Knopf, 1949.

Hoffman, W. Michael, and Jennifer Mills Moore, eds. *Ethics and the Management of Computer Technology*. Cambridge: Oelgeschlager, Gunn and Hain, 1982.

Holland, John H. *Adaptation in Natural and Artificial Systems: An Introductory Analysis with Applications to Biology, Control, and Artificial Intelligence*. Cambridge, Mass.: MIT Press, 1992.

Hollis, Martin. *The Cunning of Reason*. Cambridge and New York: Cambridge University Press, 1987.

———. *Explaining and Understanding International Relations*. Oxford and New York: Oxford University Press, 1990.

Hollis, Martin, and Steven Lukes, eds. *Rationality and Relativism*. Cambridge, Mass.: MIT Press, 1982.

Holloway, Karla F. C. *Moorings and Metaphors: Figures of Culture and Gender in Black Women's Literature*. New Brunswick, N.J.: Rutgers University Press, 1992.

Home Mortgage Disclosure Act of 1975. U.S. Code. vol. 4, secs. 2801–10 (1988).

Honderich, Ted. *Political Violence*. Ithaca, N.Y.: Cornell, 1976.

Hook, Sidney. *Philosophy and Public Policy*. Carbondale/Edwardsville: Southern Illinois University Press, 1980.

Hook, Sidney, ed. *Human Values and Economic Policy*. New York: New York University Press, 1967.

Hooker, C. A., J. J. Leach, and E. F. McClennen, eds. *Foundations and Applications of Decision Theory*. Dortrecht, Holland: Reidel, 1978.

Hopkins, Raymond F., and Donald J. Puchala, eds. *The Global Economy of Food*. Madison: University of Wisconsin Press, 1978.

Horowitz, I. L., and J. E. Katz. *Social Science and Public Policy in the U.S.* New York: Praeger, 1975.

Horton, Robin. *Patterns of Thought in Africa and the West: Essays on Magic, Religion, and Science.* Cambridge and New York: Cambridge University Press, 1993.

Hume, David. *A Treatise of Human Nature.* Oxford: Oxford University Press, 1888; 1967.

———. *Enquiry Concerning the Principles of Morals.* In D. Raphael. *British Moralists 1650–1800, II.* London: Oxford University Press, 1969.

Hysom, John L., and William J. Bolce. *Business and Its Environment.* St. Paul, Minn.: West, 1983.

Iannone, A. Pablo. "Character Traits." Ph.D. thesis. University of Wisconsin, 1975.

———. "Review of Sterling M. McMurrin, The Tanner Lectures on Human Values, VIII." *History of European Ideas* 10, no. 4 (1989): 484–85.

———. "Informing the Public: Ethics, Policy Making, and Objectivity in News Reporting." *Philosophy in Context* 20 (1990): 1–21.

———. "South." In Fernando Alegria and Jorge Ruffinelli, eds. *Paradise Lost or Gained? The Literature of Hispanic Exile.* Houston: Arte Público Press, 1990.

———. "Critical Interaction: Judgment, Decision, and the Social Testing of Moral Hypotheses." *International Journal of Social and Moral Studies* 6, no. 2 (summer 1991): 135–48.

———. "Social Traps: High-Tech Weapons, Rarefied Theories, and the World of Politics." *Epistemologia: Italian Journal for the Philosophy of Science* (1991): 219–37.

———. "Social Choice Theory: Formalism Infatuation and Policy Making Realities." *Epistemologia: Italian Journal for the Philosophy of Science* (1992): 263–78.

———. *Philosophy as Diplomacy: Essays in Ethics and Policy Making.* Amherst, N.Y.: Humanity Books, 1994.

———. "Cross-Cultural Ecologies: The Expatriate Experience, the Multiculturalism Issue, and Philosophy." In Nancy E. Snow, ed., *In the Company of Others: Perspectives on Community, Family, and Culture.* Lanham, Md., and London: Rowman & Littlefield, 1996.

———. "Dealing with Diversity: Cultural Fragmentation, Intercultural Conflicts, and Philosophy," in Robert Paul Churchill, ed., *Crossing Cultural Boundaries.* Amherst, N.Y.: Humanity Books, forthcoming.

———. Entries on the South American philosophers Carlos Astrada, Vicente Fatone, Francisco Miró Quesada Cantuarias, and Alberto Rougés. In Stuart Brown, Diané Collinson, and Robert Wilkinson, eds. *Biographical Dictionary of Twentieth-Century Philosophers.* London and New York: Routledge, 1996.

Iannone, A. Pablo, ed. *Contemporary Moral Controversies in Technology.* New York and London: Oxford University Press, 1987.

———. *Contemporary Moral Controversies in Business.* New York and London: Oxford University Press, 1989.

———. *Through Time and Culture.* Englewood Cliffs, N.J.: Prentice Hall, 1994.

Jaeger, W. *Paideia: The Ideals of Greek Culture.* New York: Oxford University Press, 1939.

James, William. *The Will to Believe and Other Essays in Popular Philosophy.* New York: Dover, 1956.

Jenkins, W. I. *Policy Analysis.* New York: St. Martin's, 1978.

Jones, Charles O. *An Introduction to the Study of Public Policy.* Belmont, Calif.: Duxbury/ Wadsworth, 1970.

Kahneman, Daniel, Paul Slovic, and Amos Tversky, eds. *Judgment Under Uncertainty: Heuristics and Biases.* Cambridge: Cambridge University Press, 1982.

Kammeyer, Kenneth C. W., ed. *Population Studies: Selected Essays and Research.* Chicago: Rand McNally, 1975.

Kanel, Don. "Institutional Economics: Perspectives on Economy and Society." *Journal of Economic Issues* 19, no. 3 (1985): 815–28.

————. "The Human Predicament: Society, Institutions, and Individuals." *Journal of Economic Issues* 22, no. 2 (1988): 427–34.

Kant, Immanuel. *On History*. Edited by Lewis White Beck. Indianapolis and New York: Bobbs-Merrill, 1963.

————. *The Metaphysical Principles of Virtue*. Translated by James Ellington. Indianapolis and New York: Bobbs-Merrill, 1964.

————. *The Metaphysical Elements of Justice*. Edited by John Ladd. Indianapolis and New York: Bobbs-Merrill, 1965.

————. *Foundations of the Metaphysics of Morals*. Edited by Robert Paul Wolff. Indianapolis and New York: Bobbs-Merrill, 1969.

Kavanagh, John P. "Ethical Issues in Plant Relocation." *Business and Professional Ethics Journal* (winter 1982): 21–33.

Kearney, Hugh. *Science and Change: 1500–1700*. New York and Toronto: McGrawHill, 1971.

Kelley, Ron, and Jonathan Friedlander. *Irangeles: Iranians in Los Angeles*. Berkeley: University of California Press, 1993.

Keynes, John Meynard. *The General Theory of Employment, Interest, and Money*. New York: Harcourt, Brace & World, 1964.

Knopman, Debra S., and Richard A. Smith. "Twenty Years of the Clean Water Act." *Environment* 35, no. 1 (January/February 1993): 17–20, 34–41.

Kohlberg, L. *The Philosophy of Moral Development*. San Francisco: Harper and Row, 1981.

Kokotovic, P. V., ed. *Foundations of Adaptive Control*. Berlin and New York: SpringerVerlag, 1991.

Kroeber, A. L. *Anthropology Today: An Encyclopedic Inventory*. Chicago: University of Chicago Press, 1953.

Kroeber, A. L., and Clyde Kluckhohn. *Culture: A Critical Review of Concepts and Definitions*. New York: Random House, 1963.

Krushwitz, Robert B., and Robert C. Roberts. *The Virtues*. Belmont, Calif.: Wadsworth, 1987.

Kruytbosch, C. E. *Academic-Corporate Research Relationships: Forms, Functions, and Fantasies*. Washington, D.C.: The National Science Foundation, 1981.

Kuehn, Thomas J., and Alan L. Porter, eds. *Science, Technology, and National Policy*. Ithaca, N.Y., and London: Cornell University Press, 1981.

Kuhn, Thomas S. *The Copernican Revolution*. Cambridge, Mass.: Harvard University Press, 1951.

————. *The Structure of Scientific Revolutions*. 2d ed. Chicago: University of Chicago Press, 1970.

Kurokawa, Kisho. *The Architecture of Symbiosis*. New York: Rizzoli, 1988.

Lacey, A. R. *A Dictionary of Philosophy*. New York: Scribner's Sons, 1976.

Lackner, Lucas. *Internal Exile: Poems & Illustrations*. Santa Barbara, Calif.: Santa Barbara Press, 1983.

Lakatos, Imre, and Alan Musgrave, eds. *Criticism and the Growth of Knowledge*. Cambridge: Cambridge University Press, 1970.

Lambert, Jacques. *Latin America: Social Structures and Political Institutions*. Berkeley and Los Angeles: University of California Press, 1969.

Lamont, W. D. *The Value Judgement*. Edinburgh, Scotland: The Edinburgh University Press, 1955.

Lanys, G. A., and W. C. McWilliams. *Crisis and Continuity in World Politics*. New York: Random House, 1966.

Lappé, F. M., et al. *Aid as Obstacle*. San Francisco: Institute for Food and Development Policy, 1980.

Larrain, J. *The Concept of Ideology*. London: Hutchinson, 1983.

Larson, Gerald James, and Eliot Deutsch. *Interpreting Across Boundaries: New Essays in Comparative Philosophy*. Princeton, N.J.: Princeton University Press, 1988.

Lecky, William Edward Hartpole. *The Map of Life*. London: Longmans, Green, 1921.

Ledooar, Robert J. *U.S. Food and Drug Multinationals in Latin America*. New York: IDOC, North America, Inc., 1975.

Lekachman, Robert. *Economists at Bay*. New York: McGraw-Hill, 1976.

Lenin, V. "What Is to Be Done?" *Selected Works*. Moscow: Progress Publishers, 1963.

Leopold, Aldo. *A Sand County Almanac and Sketches Here and There*. London, Oxford, and New York: Oxford University Press, 1949.

Levi, Isaac. *Hard Choices*. Cambridge: Cambridge University Press, 1986.

Levidow, Les, and Bob Young, eds. *Science, Technology, and the Labour Process*. Vol. 1. Atlantic Highlands, N.J.: Humanities Press, 1981.

———. *Science, Technology, and the Labour Process*. Vol. 2. Atlantic Highlands, N.J.: Humanities Press, 1985.

Lewin, Kurt. "Feedback Problems of Social Diagnosis and Action, Part II-B of Frontiers in Group Dynamics." *Human Relations* 1 (1947): 147–53.

———. *Field Theory in Social Science*. New York: Harper and Brothers, 1951.

Lieberman, Jathro K. *The Litigious Society*. New York: Basic Books, 1981.

Limerick, Patricia Nelson. "Information Overload Is a Prime Factor in Our Culture Wars." *The Chronicle of Higher Education*, July 29, 1992: A32.

Lindblom, Charles E., and David K. Cohen. *Usable Knowledge*. New Haven, Conn.: Yale University Press, 1979.

Lineberry, Robert L. *American Public Policy*. New York: Harper & Row, 1977.

Linnen, Beth, ed. "Housing Equity Funds Test the Tolerance of FIRREA Rules." *Savings Institutions* (June 1990): 12–14.

Lloyd, Dennis. *Public Policy*. London: Athlone Press, 1953.

Lockhart, James. *Spanish Peru, 1532–1560: A Colonial Society*. Madison. Milwaukee, London: University of Wisconsin Press, 1968.

Lockhart, James, and Stuart B. Schwartz. *Early Latin America: A History of Colonial Spanish America and Brazil*. Cambridge and New York: Cambridge University Press, 1983.

Louthan, William C. *The Politics of Justice: A Study of Law, Social Science, and Public Policy*. New York: Kenikat, 1979.

Lugones, Maria. "Playfulness, 'World'-Travelling, and Loving Perception." *Hypatia* 1, no. 2 (summer 1987): 3–19.

Lukacs, G. *History and Class Consciousness*. London: Merlin Press, 1971.

Luke, Martell. *Ecology and Society: An Introduction*. Cambridge: Polity Press, 1994.

Lyotard, Jean-Francois. *The Postmodern Condition: A Report on Knowledge*. Minneapolis: University of Minnesota Press, 1984.

Machan, Tibor R. *The Moral Case for a Free Market Economy: A Philosophical Argument*. Lewiston, N.Y.: E. Mellen Press, 1988.

———. *Capitalism and Individualism: Reframing the Argument for a Free Society*. New York: Harvester Wheatsheaf, 1990.

MacIntyre, Alasdair. *After Virtue*. Notre Dame, Ind.: Notre Dame University Press, 1981.

———. "Utilitarianism and Cost-Benefit Analysis: An Essay on the Relevance of Moral Philosophy and Bureaucratic Theory." In D. Scherer and T. Attia, eds. *Ethics and the Environment*. Englewood Cliffs, N.J.: Prentice-Hall, 1983.

———. "Relativism, Power, and Philosophy." In Kenneth Baynes, James Bohman, and Thomas McCarthy, eds. *After Philosophy*. Cambridge, Mass.: MIT Press, 1987): 385–411.

———. *Whose Justice? Which Rationality?* Notre Dame, Ind.: University of Notre Dame Press, 1988.

Mackie, J. L. *Inventing Right and Wrong.* New York: Penguin Books, 1977.

Mahler, Margaret S., Fred Pine, and Ann I. Bergman. *The Psychological Birth of the Human Infant: Symbiosis and Individuation.* New York: Basic Books, 1975.

Majone, Giandomenico. *Evidence, Argument, and Persuasion in the Policy Process.* New Haven, Conn.: Yale University Press, 1989.

Mandelbrot, Benoit B. *The Fractal Geometry of Nature.* New York and San Francisco: W. H. Freeman, 1984.

Manheim, Karl. *Ideology and Utopia.* London: Routledge and Kegan Paul, 1936.

Martínez Estrada, Ezequiel. *Radiografía de la Pampa.* Bueños Aires: Losada, 1953.

Marx, Karl. *Selected Writings.* Oxford: Oxford University Press, 1977.

Marx, Karl, and Friedrich Engels. "The German Ideology." In Robert C. Tucker, ed. *The Marx-Engels Reader.* New York: W. W. Norton, 1972, pp. 110–64.

Mason, Stephen F. *A History of the Sciences.* New York: Collier, 1962.

May, Judith, and Aaron B. Wildavsky. *The Policy Cycle.* Beverly Hills and London: Sage, 1978.

Mayr, Otto. *Feedback Mechanisms.* Washington, D.C.: Smithsonian Institution Press, 1971.

Mazmanian, D. A., and P. A. Sabatier. *Effective Policy Implementation.* Lexington, Mass.: Lexington Books, 1981.

McCary, J. L., ed. *Psychology of Personality.* New York and London: Grove Press and Evergreen Books, 1959.

McCloskey, H. J. *Ecological Ethics and Politics.* Totowa, N.J.: Rowman and Littlefield, 1983.

McMurrin, Sterling M., ed. *The Tanner Lectures on Human Values, VIII.* Salt Lake City, Utah, and Cambridge: University of Utah Press/Cambridge University Press, 1988.

Mead, Margaret. "The Cross-Cultural Approach to the Study of Personality." in J. L. McCary, ed. *Psychology of Personality.* New York and London: Grove Press and Evergreen Books, 1959, pp. 201–52.

Meiland, Jack W., and Michael Krausz. *Relativism: Cognitive and Moral.* Notre Dame, Ind., and London: University of Notre Dame Press, 1982.

Merchant, Carolyn, ed. *Ecology.* Amherst, N.Y.: Humanity Books, 1994.

Merton, Robert K. *Social Theory and Social Structure.* New York: Free Press, 1968.

Mesthene, Emmanuel G. *Technological Change.* Cambridge, Mass.: Harvard University Press, 1970.

Meyer, Laurence H. *Macroeconomics.* Cincinnati: SouthWestern, 1980.

Mikula, G. *Justice and Social Interaction.* New York: Springer-Verlag, 1980.

Mill, John Stuart. *Principles of Political Economy.* London: Parker, 1848.

———. "On Liberty." London: Parker, 1859. Also in Mary Warnock, ed. *Utilitarianism, On Liberty, Essay on Bentham, Together with Selected Writings of Jeremy Bentham and John Austin.* Cleveland and New York: Meridian, 1968, pp. 126–250.

———. "Utilitarianism." In Max Lerner, ed. *Essential Works of John Stuart Mill.* New York and London: Bantam Books, 1961.

———. *Principles of Political Economy, with Some of Their Applications to Social Philosophy.* 5th ed. New York: D. Appleton, 1883.

Milo, Ronald D. *Immorality.* Princeton, N.J.: Princeton University Press, 1984.

Minogue, K. *Alien Powers: The Pure Theory of Ideology.* London: Weidenfeld and Nicholson, 1985.

Mintzber, H. *Power In and Around Organizations.* Englewood Cliffs, N.J.: Prentice-Hall, 1983.

Mitcham, Carl. "Review of *Contemporary Moral Controversies in Technology.*" In *Canadian Philosophical Reviews* 7, no. 8 (August 1987): 311–13.

Monastersky, Richard. "The Light at the Bottom of the Ocean." *Science News* 150, no. 10 (September 7, 1996): 156–57.

———. "Deep Dwellers: Microbes Thrive Far Below Ground." *Science News* 151, no. 13 (March 29, 1997): 192–93.

Moessinger, Pierre. *La psychologie morale*. Paris: Presses Universitaires de France, 1989.

―――. "La théorie du choix rationnel: critique d'une explication." *Information sur les sciences sociales*. London, Newbury Park, and New Delhi: SAGE 31, no. 1, 1992, pp. 87–111.

Molander, E. *Responsive Capitalism: Case Studies in Corporate Social Conduct*. New York: McGraw-Hill, 1980.

Moos, Rudolf H., and Paul M. Insel. *Issues in Social Ecology: Human Milieus*. Palo Alto, Calif.: National Press Books, 1974.

Morner, Magnus. *Race Mixture in the History of Latin America*. Boston: Little, Brown, 1967.

Morrison, Roy. *Ecological Democracy*. Boston: South End Press, 1995.

Moser, Paul K. *Rationality in Action*. New York and Cambridge: Cambridge University Press, 1990.

Murphy, Arthur E. *Reason and the Common Good*. Englewood Cliffs, N.J.: Prentice Hall, 1963.

―――. *The Theory of Practical Reason*. La Salle, Ill.: Open Court, 1965.

Naess, Arne. *Ecology, Community, and Lifestyle: Outline of an Ecosophy*. Cambridge and New York: Cambridge University Press, 1990.

Naficy, Hamid. *The Making of Exile Cultures: Iranian Television in Los Angeles*. Minneapolis: University of Minnesota Press, 1993.

Nagel, Stuart S. *Improving Policy Analysis*. Beverly Hills and London: Sage, 1980.

Nagel, Thomas. *Mortal Questions*. Cambridge and New York: Cambridge University Press, 1979.

―――. *The View from Nowhere*. New York: Oxford University Press, 1986.

―――. "Moral Conflict and Political Legitimacy." *Philosophy and Public Affairs* 16, no. 3 (summer 1987): 215–240.

Nash, Hugh, ed. *The Energy Controversy*. San Francisco: Friends of the Earth, 1979.

Navarro, Vicente. "Class and Race: Life and Death Situations." *Monthly Review* (September 1991): 1–13.

Nelkin, Dorothy. *The Languages of Risk*. Beverly Hills and London: Sage, 1985.

Nelkin, Dorothy, ed. *Controversy: Politics of Technical Decisions*. Beverly Hills: Sage, 1979.

Newton, Lisa H., and Catherine K. Dillingham. *Watersheds*. Belmont, Calif.: Wadsworth, 1993.

Nicolson, Sir Harold. *Diplomacy*. 3d ed. London, Oxford, and New York: Oxford University Press, 1963.

Nielsen, Kai. "Philosophy as Critical Theory." *Proceedings and Addresses of the American Philosophical Association* 61, supp. (September 1987): 89–108.

Nietzsche, Friedrich. *Beyond Good and Evil*. New York: Vintage Books, 1966.

Onyewuenyi, Innocent C. "Traditional African Aesthetics: A Philosophical Perspective." *International Philosophical Quarterly* 24. no. 3 (September 1984): 237–44.

Oppenheim, Felix E. *Political Concepts: A Reconstruction*. Chicago: University of Chicago Press, 1981.

Ortega y Gasset, José. "The Dehumanization of Art." In *The Dehumanization of Art and Notes on the Novel*. Princeton, N.J.: Princeton University Press, 1948.

Oser, Jacob. *The Evolution of Economic Thought*. New York: Harcourt, Brace & World, 1963.

Palumbo, D. J., and M. A. Harder. *Implementing Public Policy*. Lexington, Mass.: Lexington Books, 1981.

Papathanasis, Anastasios, and Christopher Vasillopulos. "Task and Job: The Promise of Transactional Analysis." *American Journal of Economics and Sociology* 50, no. 2 (April 1991): 169–82.

Parker, Donn B. *Ethical Conflicts in Computer Science and Technology*. Arlington, Mass.: AFIPS, 1977.

Parsons, Talcott, Edward Shils, Kaspar D. Naegele, and Jesse R. Pitts, eds. *Theories of Society*. New York: Free Press, 1965.

Passmore, John. *Man's Responsibility for Nature*. New York: Charles Scribner's Sons, 1974.

Paul, Ellen Frankel, Fred D. Miller Jr., and Paul Jeffrey, eds. *Ethics and Economics*. Oxford: Basil Blackwell, 1985.

Peirce, Charles Sanders. "Abduction and Induction." In Justus Buchler, ed. *Philosophical Writings of Peirce*. New York: Dover Publications, 1940, pp. 150–56.

———. "The Fixation of Belief." *Popular Science Monthly* (November 1877): 1–15.

Pennock, J. Roland, and John W. Chapman. *Privacy*. New York: Atherton, 1971.

Perelman, M. *Farming for Profit in a Hungry World: Capital and the Crisis in Agriculture*. Montclair, N.J.: Allanheld, Osmun, 1977.

Perry, Richard J. "Why Do Multiculturalists Ignore Anthropologists?" *Chronicle of Higher Education*, March 4, 1992: A52.

Perry, Thomas D. *Moral Reasoning and Truth*. Oxford: Clarendon Press, 1976.

Phillips, A. W. "Mechanical Models in Economic Dynamics." *Econometrica* 17 (1950): 283–305.

———. "Stabilization Policy in a Closed Economy." *Economic Journal* (June 1954): 290–305.

———. "Stabilization Policy and the Time-Forms of Lagged Responses." *Economic Journal* (June 1957): 265–77.

Pike, Frederick B. *Latin American History: Selected Problems*. New York: Harcourt, Brace & World, 1969.

Plall, J. "Social Traps." *American Psychologist* 28 (1973): 642–51.

Plato. *The Collected Dialogues*. Princeton, N.J.: Princeton University Press, 1973.

Popper, Karl. *The Logic of Scientific Discovery*. New York: Scientific Editions, 1961.

———. *Conjectures and Refutations*. New York: Harper, 1965.

Potter, Nelson T., and Mark Timmons, eds. *Morality and Universality*. Dordrecht, Holland, and Boston: Reidel, 1986.

Press, Frank, Leon T. Silver, et al. *New Pathways in Science and Technology*. New York: Vintage, 1985.

Pytlik, E. C., D. P. Lauda, and D. L. Johnson. *Technology, Change, and Society*. Worcester, Mass.: Davis, 1978.

Quade, E. S. *Analysis for Public Decision*. Amsterdam: North Holland, 1982.

Raleigh, Walter, ed. *The Complete Works of George Savile, First Marquess of Halifax*. Oxford: Clarendon Press, 1912.

Rapoport, Eduardo H. "Vida en Extinción." *Ciencia Hoy* 2, no. 10 (November–December 1990): 26–35.

Ravitch, Diane. *The Revisionists Revised: A Critique of the Radical Attack on the Schools*. New York: Basic Books, 1978.

———. "Diversity and Democracy: Multicultural Education in America." *American Educator: The Professional Journal of the American Federation of Teachers* 14, no. 1 (spring 1990): 16–20, 46–48.

———. "Integration, Segregation, Pluralism." *American Scholar* 45, no. 2 (spring 1976): 206–17.

———. "Multiculturalism Yes, Particularism No." *Chronicle of Higher Education*, October 24, 1990: A44.

———. "A Culture in Common." *Educational Leadership* 49, no. 4 (December 1991): 8–11.

Rawls, John. *A Theory of Justice*. Cambridge, Mass.: Harvard University Press, 1971.

Regan, Richard J. *The Moral Dimensions of Politics*. New York and Oxford: Oxford University Press, 1986.

Regan, Tom, ed. *Matters of Life and Death*. New York: Random House, 1980.

————. *Earthbound*. New York: Random House, 1984.

Revkin, Andrew. *The Burning Season*. Boston: Houghton Mifflin, 1990.

Rhoads, Steven E. *The Economist's View of the World*. Cambridge and New York: Cambridge University Press, 1985.

Rich, Robert F. *Translating Evaluation into Policy*. Beverly Hills and London: Sage, 1979.

Richardson, George P. *Feedback Thought in Social Science and Systems Theory*. Philadelphia: University of Pennsylvania Press, 1991.

Roberts, K. H., and L. Burstein. *Issues in Aggregation*. San Francisco: Jossey Bass, 1980.

Rohr, John. *Ethics for Bureaucrats: An Essay on Law and Values*. New York: Dekkar, 1978.

Rose, Richard. *The Dynamic of Public Policy*. Beverly Hills and London: Sage, 1976.

Rosen, Cory. *Employee Ownership: An Alternative to Plant Closings*. Arlington, Va.: National Center for Employee Ownership, 1982.

Rosen, Stanley. "Philosophy and Revolution." *Independent Journal of Philosophy* 3 (1979): 71–78.

Roy, Rustum. "Science for Public Consumption: More than We Can Chew?" *Technology Review* (April 1983): 1–13.

————. "Technological Literacy: An Uphill Battle." *Technology Review* (November–December 1983): 18–19.

Sagoff, Mark. *The Economy of the Earth*. Cambridge: Cambridge University Press, 1988.

Sayre, Kenneth M., et al. *Regulation, Values, and the Public Interest*. Notre Dame, Ind.: University of Notre Dame Press, 1980.

Scanlon, T. M. "The Significance of Choice." In Sterling M. McMurrin. *The Tanner Lectures on Human Values, VIII*. Cambridge and Salt Lake City: Cambridge University Press/University of Utah Press, 1988, pp. 149–216.

Scheingold, Stuart A. *The Politics of Rights*. New Haven, Conn., and London: Yale University Press, 1974.

Schelling, T. C. *Micromotives and Macrobehavior*. New York: W. W. Norton, 1978.

Schieber, Paul H., and Dennis Replansky. *The Lender's Guide to Consumer Compliance and Anti-Discrimination Laws*. Chicago: Probus, 1991.

Schilpp, Paul, ed. *The Philosophy of Alfred North Whitehead*. Evanston, Ill.: Northwestern University, 1941.

Schneider, Claudine. "Preventing Climate Change." *Issues in Science and Technology* (summer 1989): 55–62.

Schroeder, D. A., and D. E. Johnson. "Utilization of Information in a Social Trap." *Psychological Reports* 50 (1982): 107–13.

Schulman, Paul R. *Large-Scale Policy Making*. New York: Elsevier, 1980.

Schumacher, E. F. *Small Is Beautiful: Economics as If People Mattered*. New York: Harper & Row, 1973.

Schumpeter, Joseph A. *Capitalism, Socialism, and Democracy*. New York and Evanston, Ill.: Harper & Row, 1942.

Schuurman, Egbert. *Technology and the Future*. Toronto: Wedge, 1980.

Schwartz, Thomas. *The Logic of Collective Choice*. New York: Columbia University Press, 1986.

Schweitzer, Glenn E. *Techno-Diplomacy*. New York and London: Plenum Press, 1989.

Scioli, F. P., Jr., and Thomas J. Cook. *Methodologies for Analyzing Public Policies*. Lexington, Mass.: Lexington Books, 1975.

Seaborg, G. T. "Science. Technology, and Development: A New World Outlook." *Science* 181 (1973): 13–19.

Searle, John R. *Speech Acts*. New York and London: Cambridge University Press, 1974.

Self, Peter. *Econocrats and Policy Process: The Politics and Philosophy of Cost-Benefit Analysis*. London: Macmillan, 1975.

Seliger, Martin. *Ideology and Politics*. New York: Free Press, 1976.

———. *The Marxist Conception of Ideology: A Critical Essay*. Cambridge and New York: Cambridge University Press, 1977.

Sellers, Leonard L., and William L. Rivers, eds. *Mass Media Issues*. Englewood Cliffs, N.J.: Prentice-Hall, 1977.

Sem, *Tangoville sur mer: Every body is doing it now!* Paris: Success, 1913.

Sen, Amartya. "The Moral Standing of the Market." *Social Philosophy and Policy* 22 (spring 1985): 1–19.

———. "Rational Fools: A Critique of the Behavioral Foundations of Economic Theory." *Philosophy and Public Affairs* 6 (summer 1977): 317–44.

Sen, Amartya, and Bernard Williams, eds. *Utilitarianism and Beyond*. Cambridge and New York: Cambridge University Press, 1982.

Shackle, G. L. S. *Time and Economics*. Amsterdam: North Holland, 1958.

Shain, Yossi. *The Frontier of Loyalty: Political Exiles in the Age of the Nation-State*. Middletown, Conn.: Wesleyan University Press, 1989.

———. *Governments in Exile in Contemporary World Politics*. New York: Routledge, 1991.

Shapiro, Michael. *Language and Politics*. New York: New York University Press, 1984.

Shea, Christopher. "Who Wars with Whom?" *Chronicle of Higher Education* (July 5, 1996): A6, A7, A12.

Sherwin, Douglas S. "The Ethical Roots of the Business System." *Harvard Business Review* (November–December 1983): 183–87, 189–92.

Shklar, Judith. "Decisionism." In C. J. Friedrich, ed. *Nomos*. Vol. 7. *Rational Decision*. New York: Atherton, 1964, pp. 3–17.

Shlav, Ann B. "Financing Community: Methods for Assessing Residential Credit Disparities, Market Barriers, and Institutional Reinvestment Performance in the Metropolis." *Journal of Urban Affairs* 11, no. 3 (1989): 201–23.

Shournatoff, Alex. *The World Is Burning*. Boston: Little, Brown, 1990.

Shrader-Frechette, K. *Risk Analysis and Scientific Method*. Dordrecht, Holland, and Boston: D. Reidel, 1985.

———. *Science Policy, Ethics, and Economic Methodology*. Dordrecht, Holland, and Boston: D. Reidel, 1985.

Sidgwick, Henry. *The Methods of Ethics*. London: Macmillan, 1874.

———. *The Elements of Politics*. London: Macmillan, 1891.

———. *Practical Ethics*. London: Macmillan, 1898.

Sikora, R., and B. Barry, eds. *Obligations to Future Generations*. Philadelphia: Temple University Press, 1978.

Simmel, Georg. *The Web of Group-Affiliations* New York: Free Press, 1964.

———. *On Individuality and Social Norms*. Chicago: Chicago University Press, 1971.

Simon, Herbert. *Administrative Behavior: A Study of Decision-Making Processes in Administrative Organization*. New York: Macmillan, 1947.

———. "On the Application of Servomechanism Theory in the Study of Production Control." *Econometrica* 20, no. 2 (1952): 247–68.

———. "Some Strategic Considerations in the Construction of Social Science Models." In P. Lazarfeld, ed. *Mathematical Thinking in the Social Sciences*. Glencoe, Ill.: Free Press, 1954.

———. *Models of Man*. New York: Wiley, 1961.

———. "The Architecture of Complexity." *Proceedings of the American Philosophical Society* 106, no. 6 (1962): 467–82.

———. "Rational Decision Making in Business Organizations." *American Economic Review* 69, no. 4 (1978): 493–513.

Simon, Yves R. *The Definition of Moral Virtue.* New York: Fordham University Press, 1986.

Singer, Marcus G. "Freedom from Reason." *Philosophical Review* 89 (April 1970): 253–61.

———. "Is Ethics a Science? Ought It to Be?" *Zygon* 15, no. 1 (March 1980): 37–38.

———. "Moral Issues and Social Problems: The Moral Relevance of Moral Philosophy." *Philosophy* 60 (1985): 10–16.

———. "The Ideal of a Rational Morality." *American Philosophical Association's Proceedings and Addresses* 60, no. 1 (September 1986): 15–38.

———. "Some Preliminary Observations on Truth in Ethics." *Philosophy in Context* 16 (1986), Moral Truth: 11–6.

Singer, Marcus George. *Generalization in Ethics.* New York: Atheneum, 1961.

Singer, Marcus G., ed. *Morals and Values.* New York: Charles Scribner's Sons, 1977.

Skindrud, E. "Romanian Cave Contains Novel Ecosystem." *Science News* 149 (June 29, 1996): 405.

Slaughter, John B., et al. *Emerging Issues in Science and Technology,* 1981 Washington, D.C.: National Science Foundation, 1982.

Smart, C. E., and W. T. Stanbury. *Studies on Crisis Management.* Toronto: Institute for Research on Public Policy, 1978.

Smart, J. J. C., and Bernard Williams. *Utilitarianism: For and Against.* Boston: Beacon Press, 1957.

Smith, Adam. *An Inquiry into the Nature and Causes of the Wealth of Nations.* Oxford: Clarendon Press, 1976.

Smith, D. C., and A. E. Douglas. *The Biology of Symbiosis.* London and Baltimore, Md.: E. Arnold, 1987.

Smith, John E. *The Spirit of American Philosophy.* New York: Oxford University Press, 1966.

Socolow, Robert H. "Failures of Discourse: Obstacles to Integration of Environmental Values into Natural Resource Policy." In D. Scherer and T. Attig, eds. *Ethics and the Environment.* Englewood Cliffs, N.J.: Prentice-Hall, 1983, pp. 152–69.

Spencer, Herbert. *Social Statics or Order Abridged and Revised.* New York: Appleton, 1915.

Spiro, Melford E. *Oedipus in the Trobriands: The Making of a Scientific Myth.* Chicago: University of Chicago Press, 1982.

———. "Some Reflections on Cultural Determinism and Relativism with Special Reference to Emotion and Affection." In Richard A. Shweder and Robert A. LeVine. *Culture Theory: Essays on Mind, Self, and Emotion.* Cambridge and New York: Cambridge University Press, 1984, pp. 323–46.

Staff, "GAO Says Revised Regulations May Not Resolve CRA Problems." *Housing and Development Reporter,* December 4, 1995, p. 457.

Stanlis, Peter J. *Edmund Burke and the Natural Law.* Ann Arbor: University of Michigan Press, 1958.

Star, Susan Leigh, ed. *Ecologies of Knowledge: Work and Politics in Science and Technology.* Albany: State University of New York Press, 1995.

Starleng, G., and O. W. Bashkin. *Issues in Business and Society: Capitalism and Public Purpose.* Boston: Kent, 1985.

Starling, Grover. *The Changing Environment of Business.* Boston: Kent, 1980.

Steiner, George A., and John F. Steiner. *Business, Government, and Society.* 3d ed. New York: Random House, 1980.

Stephen, James Fitzjames. *Liberty, Equality, Fraternity.* Cambridge: Cambridge University Press, 1967.

Stokey, Dith, and Richard Zeckhauser. *A Primer for Policy Analysis.* New York: W. W. Norton, 1978.

Stover, Carl F., ed. *The Technological Order*. Detroit: Wayne University Press, 1963.

Sullivan, William M. *Reconstructing Public Philosophy*. Berkeley, Los Angeles, and London: University of California Press, 1986.

Sutherland, John W. *Societal Systems*. New York: North Holland, 1978.

Sutton, Constance R., and Elsa M. Chaney. *Caribbean Life in New York City: Sociocultural Dimensions*. New York: Center for Migration Studies, 1992.

Tawney, R. H. *The Acquisitive Society*. New York: Harcourt, Brace & World, 1920.

——. *Religion and the Rise of Capitalism*. New York: Mentor, 1958.

Taylor, Charles. *Multiculturalism and "The Politics of Recognition*. Princeton, N.J.: Princeton University Press, 1992.

Torres-Saillant, Silvio, ed. *Hispanic Immigrant Writers*. New York: Ollantay Press, 1989.

Tribe, Lawrence. *Where Values Conflict: Essays on Environmental Analysis, Discourse, and Decision*. Cambridge, Mass.: Ballinger, 1976.

Trivers, Robert L. "The Evolution of Reciprocal Altruism." *The Quarterly Review of Biology* (March 1971): 35–56.

Tustin, Arnold. *The Mechanics of Economic Systems*. Cambridge, Mass.: Harvard University Press, 1953.

Tyllock, G., and R. E. Wagner. *Policy Analysis and Deductive Reasoning*. Lexington, Mass.: Lexington Books, 1978.

U.S. General Accounting Office. "Community Reinvestment Act: Challenges Remain to Successfully Implement CRA." GAO/GGD-96-23. Gaithersburg, Md.: GAO, 1995.

U.S. Government. "Hearing Before the Subcommittee on Banking, Housing, and Urban Affairs-United States Senate." 102d Cong., 2d sess., September 15, 1992. In *Current Status of the Community Reinvestment Act*. Washington, D.C.: U.S. Government Printing Office, 1992.

VanDeVeer, Donald, and Christine Pierce, eds. *People, Penguins, and Plastic Trees*. Belmont, Calif.: Wadsworth, 1986.

Volti, Rudi. *Society and Technological Change*. New York: St. Martin's, 1988.

Von Bertalaffy, Ludwig. "Toward a Physical Theory of Organic Teleology, Feedback, and Dynamics." *Human Biology* 23, no. 6 (1951): 346–61.

Wallace, James D. *Moral Relevance and Moral Conflict*. Ithaca, N.Y., and London: Cornell University Press, 1988.

Walsh, John. "Merck Donates Drug for River Blindness." *Science* 238 (October 30, 1987): 610.

Walton, C. C., ed. *The Ethics of Corporate Conduct*. Englewood Cliffs, N.J.: Prentice-Hall, 1977.

Warnock, J. G. *The Object of Morality*. London: Methuen, 1971.

Waters, James A. "Catch 20.5: Corporate Morality as an Organizational Phenomenon." *Organizational Dynamics* (spring 1978): 3–19.

Weinberg, Julius. *A Short History of Medieval Philosophy*. Princeton, N.J.: Princeton University Press, 1964.

Weiner, Norbert. *Cybernetics: Or Control and Communication in the Animal and the Machine*. Cambridge, Mass.: MIT Press, 1948.

——. *Extrapolation, Interpolation, and Smoothing of Stationary Time Series*. New York: John Wiley, 1949.

——. *The Human Use of Human Beings: Cybernetics and Society*. New York: Houghton Mifflin, 1950.

Werhane, Patricia. *Adam Smith and His Legacy for Modern Capitalism*. New York and Oxford: Oxford University Press, 1991.

Westfall, Richard S. *The Construction of Modern Science: Mechanisms and Mechanics*. New York and London: John Wiley and Sons, 1971.

Westin, Alan F. *Privacy and Freedom*. New York: Atheneum, 1967.

———. *Whistle-Blowing!* New York: McGraw-Hill, 1981.

White, David. "When Is Los Angeles Going to Grow Up?" *The Dispatch* 23, no. 1 (January 1992): 1, 3.

Whyte, William F. *Street Corner Society*. Chicago: University of Chicago Press, 1955.

Wiener, Philip P., ed. *Charles Sanders Peirce: Selected Writings*. New York: Dover, 1958.

Wildavsky, Aaron. *Searching for Safety*. New Brunswick, N.J., and Oxford: Transactions, 1989.

Wiletz, Gay. *Binding Cultures: Black Women Writers in Africa and the Diaspora*. Bloomington: Indiana University Press, 1992.

Williams, Bernard. *Moral Luck*. Cambridge: Cambridge University Press, 1981.

———. *Ethics and the Limits of Philosophy*. Cambridge, Mass.: Harvard University Press, 1985.

Williams, L. Pearce. "Normal Science, Scientific Revolutions and the History of Science." In Imre Lakatos and Alan Musgrave. *Criticism and the Growth of Knowledge*. London: Cambridge University Press, 1970.

———. *Michael Faraday*. New York: Simon & Schuster, 1971.

Williams, Robin. "Some Further Comments on Chronic Controversy." *American Journal of Sociology* (May 1966): 717–21.

Wilson, Edward O., ed. *Biodiversity*. Washington, D.C.: National Academy Press, 1988.

———. *The Diversity of Life*. Cambridge, Mass.: Belknap Press of Harvard University Press, 1992.

Wimsatt, W. K., Jr., and Monroe C. Beardsley. "The Intentional Fallacy." In W. K. Wimsatt. *The Verbal Icon*. Lexington: University of Kentucky Press, 1967.

Winner, Langdon. *Autonomous Technology*. Cambridge, Mass.: MIT Press, 1977.

Woodgate, H. S. *Planning by Network*. London: Business Books, 1977.

York, C. M. "Steps Toward a National Policy for Academic Science." *Science* 172 (1971): 29–35.

Index

abortion: morals, and senses of the term ethics, 185–87

adaptation(s): in the expatriate experience, xii, 35–39; between business, government, and the nonprofit sector, 93–94; social ecologies, and feedback systems, 107–108

adversarial: approaches and court-case overload, 98; approaches and reluctant obedience, 98; approaches and settlements parasitic on them, 84–85; approaches and the Community Reinvestment Act, 82–84, 88; approaches, of limited use in technology-business decisions, 95; approaches to environmental issues, 71

advocacy: when advisable, 93–94

aesthetic: appreciation and judgment, how objective, 133, 135–37, 143; concerns and motor-vehicle policy making, 123; concerns and zoning, 114; dialogue and social ecologies, 143–44; development, xi; experiences and a creation's being a work of art, 134–37; fragmentation, critical scrutiny, and good reasons, 143; value and cognition, 132–33; value and experience, 132–33; value across cultures, 137; value, variously functional, 133–39

aesthetics: and philosophical problems

and issues treated in this book, xii; critical scrutiny, and good reasons, 143; cross-cultural, 133; functionalism in, criticisms, 135–37; when its dialogue becomes unstable or sterile, 143

aims: characteristically at issue among significant social problems, 195

anomalies: and undercurrents in science, 151, 156

anthropology: its role in dealing with the multiculturalism issue, 21

applicability: of philosophical theories to policy making and other practical situations, one of these book's concerns, xii

arbitration: and interactions in business, 103; and nongovernmental approaches, 98; and technology policy making, 68; and the multiculturalism issue, 17; and the Community Reinvestment Act, 85; and technology-business, decisions, 95; relevant in controversies, 178

art: and cross-cultural interpretations, 133; and dialogue, 133–34; and social ecologies, 133; and this book's account, 131; how it advances, 144; how its nature is settled, 133; works, how their nature is settled, 133

artistic traditions: and our times'